AF757802

SPRINGER PROTOCOLS HANDBOOKS

For further volumes:
http://www.springer.com/series/8623

Springer Protocols Handbooks collects a diverse range of step-by-step laboratory methods and protocols from across the life and biomedical sciences. Each protocol is provided in the Springer Protocol format: readily-reproducible in a step-by-step fashion. Each protocol opens with an introductory overview, a list of the materials and reagents needed to complete the experiment, and is followed by a detailed procedure supported by a helpful notes section offering tips and tricks of the trade as well as troubleshooting advice. With a focus on large comprehensive protocol collections and an international authorship, Springer Protocols Handbooks are a valuable addition to the laboratory.

Chlorella Genomics, Cultivation, and Bioenterprise

Green Economy through Algal Innovations

Edited by

Dhanasekaran Dharumadurai

Department of Microbiology, Bharathidasan University, Tiruchirappalli, Tamil Nadu, India

Humana Press

Editor
Dhanasekaran Dharumadurai
Department of Microbiology
Bharathidasan University
Tiruchirappalli, Tamil Nadu, India

ISSN 1949-2448 ISSN 1949-2456 (electronic)
Springer Protocols Handbooks
ISBN 978-1-0716-5275-6 ISBN 978-1-0716-5276-3 (eBook)
https://doi.org/10.1007/978-1-0716-5276-3

© The Editor(s) (if applicable) and The Author(s), under exclusive license to Springer Science+Business Media, LLC, part of Springer Nature 2026

This work is subject to copyright. All rights are solely and exclusively licensed by the Publisher, whether the whole or part of the material is concerned, specifically the rights of translation, reprinting, reuse of illustrations, recitation, broadcasting, reproduction on microfilms or in any other physical way, and transmission or information storage and retrieval, electronic adaptation, computer software, or by similar or dissimilar methodology now known or hereafter developed.
The use of general descriptive names, registered names, trademarks, service marks, etc. in this publication does not imply, even in the absence of a specific statement, that such names are exempt from the relevant protective laws and regulations and therefore free for general use.
The publisher, the authors and the editors are safe to assume that the advice and information in this book are believed to be true and accurate at the date of publication. Neither the publisher nor the authors or the editors give a warranty, expressed or implied, with respect to the material contained herein or for any errors or omissions that may have been made. The publisher remains neutral with regard to jurisdictional claims in published maps and institutional affiliations.

This Humana imprint is published by the registered company Springer Science+Business Media, LLC, part of Springer Nature.
The registered company address is: 1 New York Plaza, New York, NY 10004, U.S.A.

If disposing of this product, please recycle the paper.

Preface

The book entitled *Chlorella Genomics, Cultivation and Bioenterprise: Green Economy through Algal Innovations* is a manual that illustrates the techniques involved in Isolation, Purification of axenic *Chlorella* spp. from natural water, morphological Identification using microscopic morphology, molecular characterization and identification of *Chlorella* species, including 18S rRNA gene sequencing, phylogenetic analysis, whole genome analysis of *Chlorella* using Illumina Hiseq, genome characterization using GeSeq, and metabolic pathway analysis using KEGG. Further, it describes preparation of starter culture, scaling up of *Chlorella* spp. from lab to pilot scale, large-scale cultivation in open pond, photobioreactor systems, and monitoring of growth.

The most relevant target audiences belong to the Degree in B.Sc., M.Sc., Ph.D., Microbiology, Botany, Algal technology, Food biotechnology, Microbial biotechnology, Agriculture. The secondary audiences are faculty members, researchers, and food industrial people who can use the protocol for standardizing and production of microalgae on a commercial basis and are the targeted readers of this manual.

The book consists of five broad sections under which different titles are prepared based on the total content of 40 protocols. It includes the main division of microalgae sample collection and pure culture methods, taxonomic characterization and identification of *Chlorella* species, whole genome analysis, mass cultivation and value-added products from *Chlorella* spp., scaling up from lab to pilot scale, large-scale cultivation in open pond systems, photobioreactors, harvesting techniques of *Chlorella* biomass using filtration and sedimentation, and post-harvest processing of *Chlorella* biomass by drying and powdering.

This edited Protocol is contributed by authorities across the world, including China, Hungary, India, Korea, and the United Arab Emirates. This manual help all the essentials of researchers interested to work on uses of fish feed formulation with *Chlorella* wet paste form (live/concentrated feed), production of *Chlorella* dry powder as a value-added supplemental feed, *Chlorella*-enriched feed pellet preparation, nutritional analysis of *Chlorella* aquafeed, *Chlorella* aquafeed packing, labeling, experimental validation of *Chlorella*-enriched feed in tilapia fish, preparation of dietary supplements and functional foods, preparation of polyunsaturated fatty acids (PUFAs)-omega-3 and omega-6 fatty acids, preparation and quality assessment of *Chlorella* green smoothie, preparation and nutritional and sensory analysis of *Chlorella*-based cookies, microencapsulation of *Chlorella* for sensory improvement, and preparation of bio-coating from *Chlorella* to increase shelf life of commercial fruits. It covers the bioenergy and biopolymer production, preparation of biodiesel, bioethanol, and biohydrogen from *Chlorella* biomass.

I am extremely thankful to all the authors who contributed chapters for their prompt and timely responses. I extend my earnest appreciation to Monica Suchy, Springer Nature, and their team for their constant encouragement and help in bringing out the volume in its present form. I am also indebted to Springer Nature, the Authorities of Bharathidasan

University, Tiruchirappalli, Tamil Nadu, India; sincere thanks to the Tamil Nadu State Council for Science and Technology, Chennai (TNSCST/DIT/MK/2023–24/19 dated 30.04.2024), and Anusandhan National Research Foundation (ANRF), New Delhi-110,016 (SSY/2025/002489 dated 18.12.2025), for their support in the task of publishing the book.

Tiruchirappalli, Tamil Nadu, India ***Dhanasekaran Dharumadurai***

Contents

Contributors

ANANDHI ANANDARAJ • *Centre for Drug Discovery and Development, Sathyabama Institute of Science and Technology, Chennai, Tamil Nadu, India*
FAWZI BANAT • *Department of Chemical and Petroleum Engineering, SAN Campus, Khalifa University of Science and Technology, Abu Dhabi, United Arab Emirates; Food Security and Technology Center (FSTC), SAN Campus, Khalifa University, Abu Dhabi, United Arab Emirates*
KALIYAN BARATHIKANNAN • *Department of Chemical and Petroleum Engineering, SAN Campus, Khalifa University of Science and Technology, Abu Dhabi, United Arab Emirates; Food Security and Technology Center (FSTC), SAN Campus, Khalifa University, Abu Dhabi, United Arab Emirates*
HONGWEI CHAI • *Lanzhou University, Lanzhou, China*
RAMACHANDRAN CHELLIAH • *Department of Chemical and Petroleum Engineering, SAN Campus, Khalifa University of Science and Technology, Abu Dhabi, United Arab Emirates; Food Security and Technology Center (FSTC), SAN Campus, Khalifa University, Abu Dhabi, United Arab Emirates; Department of Food Science and Biotechnology, College of Agriculture and Life Sciences, Kangwon National University, Chuncheon, Gangwon-do, Korea*
PENG CHEN • *Lanzhou University, Lanzhou, China*
DHANASEKARAN DHARUMADURAI • *Department of Microbiology, School of Life Sciences, Bharathidasan University, Tiruchirappalli, Tamil Nadu, India; National Repository for Microalgae and Cyanobacteria – Freshwater and Marine (NRMC-F&M) (Sponsored by the DBT, Govt. of India), Bharathidasan University, Tiruchirappalli, Tamil Nadu, India*
AKILAN ELUMALAI • *HUN-REN Biological Research Centre, Szeged, Hungary*
MANO GOVINDHARAJ • *Department of Chemical and Petroleum Engineering, SAN Campus, Khalifa University of Science and Technology, Abu Dhabi, United Arab Emirates; Food Security and Technology Center (FSTC), SAN Campus, Khalifa University, Abu Dhabi, United Arab Emirates*
G. HARITA RAO • *KLE Independent PU College, Bengaluru, India*
D. JAYANTHI • *Mount Carmel College, Autonomous, Bengaluru, India*
B. R. KARTHIK • *Government First Grade College, Chikkaballapura, India*
DHINESH KUMAR • *Department of Environmental Biotechnology, Tiruchirappalli, Tamil Nadu, India*
RAMAKRISHNAN GOVINDHARAJ KUMAR • *Department of Microbiology, School of Life Sciences, Tiruchirappalli, Tamil Nadu, India*
SHRIKA KUMARESAN • *Department of Microbiology, Bharathidasan University, Tiruchirappalli, Tamil Nadu, India*
YINGJIE LIU • *Lanzhou University, Lanzhou, China*
A. MARTIN PAUL • *St Joseph's University, Bengaluru, India*
S. J. MEGHANA • *Sindhi PU College, Bengaluru, India*
SRINIVAS METTU • *Department of Chemical and Petroleum Engineering, SAN Campus, Khalifa University of Science and Technology, Abu Dhabi, United Arab Emirates; Food Security and Technology Center (FSTC), SAN Campus, Khalifa University, Abu Dhabi, United Arab Emirates*

MOHAMED IMRAN MOHAMEDYOUSUFF • *P.G. and Research Department of Biotechnology, Microbiology and Bioinformatics, National College (Autonomous), Tiruchirappalli, Tamil Nadu, India*

GANGATHARAN MURALITHARAN • *National Repository for Microalgae and Cyanobacteria – Freshwater and Marine (NRMC-F&M), Bharathidasan University, Tiruchirappalli, Tamil Nadu, India; Department of Microbiology, Bharathidasan University, Tiruchirappalli, Tamil Nadu, India*

REEHANA NAZAR • *P.G and Research Department of Microbiology, Jamal Mohamed College (Autonomous) [Affiliated to Bharathidasan University], Tiruchirappalli, Tamil Nadu, India*

GOWTHAM NEETHIMOHAN • *Department of Microbiology, School of Life Sciences, Bharathidasan University, Tiruchirappalli, Tamil Nadu, India*

THAJUDDIN NOORUDDIN • *B.S. Abdur Rahman Crescent Institute of Science & Technology (Deemed to be University), Chennai, Tamil Nadu, India; National Repository for Microalgae and Cyanobacteria – Freshwater (NRMC-F) (Sponsored by the DBT, Govt. of India), Bharathidasan University, Tiruchirappalli, Tamil Nadu, India*

DEOG-HWAN OH • *Department of Food Science and Biotechnology, College of Agriculture and Life Sciences, Kangwon National University, Chuncheon, Gangwon-do, South Korea*

MANIDEEP PABBA • *Department of Chemical and Petroleum Engineering, SAN Campus, Khalifa University of Science and Technology, Abu Dhabi, United Arab Emirates; Food Security and Technology Center (FSTC), SAN Campus, Khalifa University, Abu Dhabi, United Arab Emirates*

PRAKASH PALANIVEL • *National Repository for Microalgae and Cyanobacteria – Freshwater and Marine (NRMC-F&M), Bharathidasan University, Tiruchirappalli, Tamil Nadu, India; Department of Microbiology, Bharathidasan University, Tiruchirappalli, Tamil Nadu, India*

SUBRAMANIYAN PASUPATHI • *Department of Microbiology, School of Life Sciences, Bharathidasan University, Tiruchirappalli, Tamil Nadu, India; National Repository for Microalgae and Cyanobacteria – Freshwater (NRMC-F), Bharathidasan University, Tiruchirappalli, Tamil Nadu, India*

GUHANRAJ RADHAMANALAN • *Department of Microbiology, Saveetha Dental College, Chennai, Tamil Nadu, India*

PRIYANKA JAYAM RAJENDRAN • *Department of Microbiology, School of Life Sciences, Bharathidasan University, Tiruchirappalli, Tamil Nadu, India; National Repository for Microalgae and Cyanobacteria – Freshwater (NRMC-F), Bharathidasan University, Tiruchirappalli, Tamil Nadu, India*

VENGATESAN M. RANGARAJ • *Research and Innovation Center for Graphene and Two-Dimensional Materials (RIC-2D), SAN Campus, Khalifa University, Abu Dhabi, United Arab Emirates*

ASRAF SITHIKKA RASHEED • *School of Life Sciences, Bharathidasan University, Tiruchirappalli, Tamil Nadu, India*

MUTHU ROOBINI RATHINAM • *Department of Microbiology, School of Life Sciences, Bharathidasan University, Tiruchirappalli, Tamil Nadu, India*

OMM BHUSAN SARANGI • *School of Life Sciences, Bharathidasan University, Tiruchirappalli, Tamil Nadu, India*

BHUVANESHWARI SRINIVASAN • *Department of Microbiology, School of Life Sciences, Bharathidasan University, Tiruchirappalli, Tamil Nadu, India*

SHREYA SUBRAMANYA • *Allen Digital, Bengaluru, India*

SAI KUMAR TAMMINA • *Department of Chemical and Petroleum Engineering, SAN Campus, Khalifa University of Science and Technology, Abu Dhabi, United Arab Emirates; Food Security and Technology Center (FSTC), SAN Campus, Khalifa University, Abu Dhabi, United Arab Emirates*

FAYAAZUDDIN THAJUDDIN • *National Repository for Microalgae and Cyanobacteria – Freshwater (NRMC-F) (Sponsored by the DBT, Govt. of India), Bharathidasan University, Tiruchirappalli, Tamil Nadu, India; Department of Microbiology, Bharathidasan University, Tiruchirappalli, Tamil Nadu, India*

SHAKENA FATHIMA THAJUDDIN • *National Repository for Microalgae and Cyanobacteria – Freshwater (NRMC-F) (Sponsored by the DBT, Govt. of India), Bharathidasan University, Tiruchirappalli, Tamil Nadu, India*

RAJASEKAR THIRUNAVUKKARASU • *Centre for Drug Discovery and Development, Sathyabama Institute of Science and Technology, Chennai, Tamil Nadu, India*

GOKUL VARUTHARAJ • *Department of Microbiology, Bharathidasan University, Tiruchirappalli, Tamil Nadu, India*

VISHNU RAJA VIJAYAKUMAR • *Department of Microbiology, Srimad Andavan Arts and Science College (Autonomous), Affiliated to Bharathidasan University, Tiruchirappalli, Tamil Nadu, India*

NING XU • *Lanzhou University, Lanzhou, China*

Part I

Microalgae Sample Collection and Pure Culture Methods

Chapter 1

Collection and Initial Analysis of Water Samples Potentially Containing *Chlorella* spp. from a Freshwater Environment

Prakash Palanivel, Fayaazuddin Thajuddin, Dhanasekaran Dharumadurai, and Gangatharan Muralitharan

Abstract

Chlorella is a single-celled green microalga that habitats in freshwater ecosystem. It is a key primary producer in the field of ecology and biotechnology. However, collecting samples from nature is challenging due to its tiny size and uneven distribution. This chapter discusses a standardized methodology to improve the accuracy and reliability of *Chlorella* samples obtained from natural environments by combining direct water sampling with plankton net collection. Direct sampling offers accurate data on phytoplankton communities at specific depths. Plankton nets effectively concentrate low-density populations, producing sufficient biomass for microscopic identification and culture establishment. The chapter outlines a detailed protocol for preparing, collecting, preserving, and examining them in the laboratory. Further this approach suggested an increase in the precision, reliability, and efficiency of collecting *Chlorella* samples.

Key words *Chlorella*, Freshwater microalgae, Phytoplankton sampling, Direct water sampling, Plankton net collection, Sampling methodology, Algal biodiversity

1 Introduction

Chlorella is a circular, unicellular, nonmotile green algae classified as Chlorophyta [1]. These bacteria live in a variety of freshwater environments around the world, including ponds, lakes, rivers, and seasonal water bodies. As primary producers in aquatic food webs, they are critical to ecological balance. Photosynthesis converts carbon dioxide and solar energy into organic matter [2]. As a result, their economic and scientific value is significant. *Chlorella* is required for the generation of biofuels, high-value nutraceuticals such as omega-3 fatty acids and carotenoids, protein for aquaculture, and human nutrition in biotechnology [3]. Additionally, it serves as an excellent model organism for physiological and genetic study due to its quick growth and simple cellular structure [1].

Dhanasekaran Dharumadurai (ed.), *Chlorella Genomics, Cultivation, and Bioenterprise: Green Economy through Algal Innovations*, Springer Protocols Handbooks, https://doi.org/10.1007/978-1-0716-5276-3_1,
© The Author(s), under exclusive license to Springer Science+Business Media, LLC, part of Springer Nature 2026

The accurate and complete collection of *Chlorella* from its natural habitat is essential for ecological study and biotechnological applications. However, the same qualities that make *Chlorella* so popular also provide unique sampling issues. They may be unreported or under-studied due to their small cellular size, which is typically 2 to 10 micrometers in diameter and invisible to the naked eye [4]. Furthermore, their distribution within water bodies is usually uneven or patchy due to changes in light, nutrients, temperature, and water current gradients [5, 6]. As a result, a random or non-standardized sampling technique is likely to create a non-representative sample or false negatives, undermining the credibility of future studies.

To address these problems, this methodology describes two standardized and complimentary procedures that aim to increase the potential of collecting valuable and useful *Chlorella* samples. The first technique is direct water sampling, which is typically done with Niskin or Van Dorn bottles. This method retains the natural density and composition of the phytoplankton community at that location by collecting a precise volume of water from a certain depth [6]. This method is critical for ecological study that aims to quantify cell abundance (cells per liter) or investigate vertical stratification. The second approach is plankton net collection, which involves drawing a fine-mesh conical net through water. This approach filters a large volume of suspended phytoplankton, including *Chlorella*. This approach captures *Chlorella* and other suspended phytoplankton by filtering large amounts of water, which are commonly measured in cubic meters. It is especially effective when working with tiny populations or generating enough biomass to start laboratory cultures [4].

Strategically combining both strategies can result in a more successful sampling approach. The net sample collects and prioritizes even low-density populations for identification, whereas the direct sample provides a quantitative assessment of the in-situ conditions. This protocol's main objective is to offer a straightforward, reliable, and standardized procedure for gathering *Chlorella* samples, from the first field collection to the initial laboratory validation (Fig. 1). Researchers can gather high-quality samples for a variety of objectives, including microscopic identification, species isolation, culture setup, and molecular analysis.

2 Materials

1. Personal flotation device (PFD).
2. Personal protective equipment (PPE): Gloves, safety glasses, closed-toe shoes.
3. Waterproof labels and permanent marker.

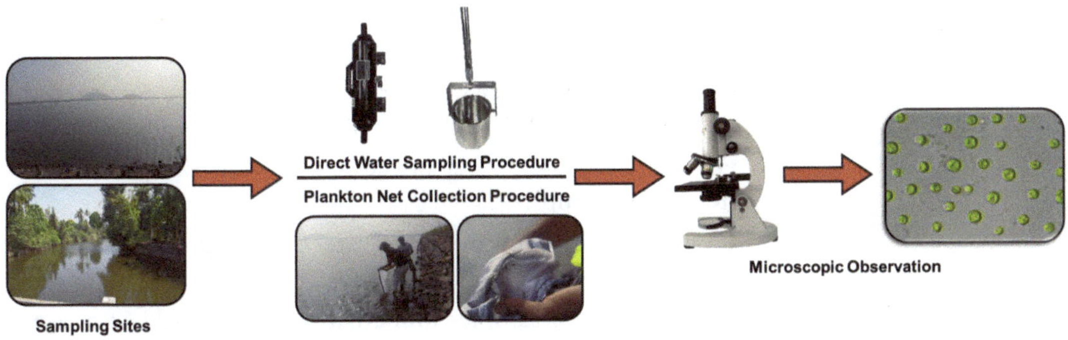

Fig. 1 Workflow for field sampling, microalgae collection, and microscopic observation

4. Field notebook and pen.
5. Cooler with ice packs.
6. Distilled water (in a rinse bottle).
7. 70% ethanol (in a spray bottle).
8. Water sampler (Van Dorn, Niskin, or Kemmerer bottle; or a telescopic pole sampler for shallow water).
9. Sample containers: Sterile, Amber Wide Mouth Bottle (HDPE) or glass bottles (500 mL to 1 L).
10. Thermometer.
11. Portable pH Meter.
12. Refractometer.
13. Conical net with a 30-cm ring diameter and a mesh size of 10–20 μm.
14. Collection Vial (Cod-End): A removable, sturdy plastic vial (50–100 mL) attached to the end of the net.
15. Plastic sample vial (10–30 mL).

3 Methods

3.1 Presampling Preparation

1. Samples are collected from an appropriate freshwater sampling sites and the exact GPS coordinates are recorded.
2. Rinse all samplers, bottles, nets, and plastic sample vials thoroughly using distilled water.
3. Sterilize sample containers and plastic vials with 70% ethanol and allow to air dry.
4. Label all the sample containers with a unique ID, date, location, and sampling method.

3.2 Direct Water Sampling Procedure

1. Record the date, time, GPS coordinates, water temperature, pH/conductivity, and weather conditions at the site.
2. Lower the open water sampler (e.g., a Van Dorn bottle) to the required depth (0.5 m) and activate the closure mechanism.
3. For shallow water, fill a bottle at the desired depth using a sterile pole sampler.
4. Carefully pour the collected water into a prelabeled, sterile container until it is nearly full.
5. Place one set of samples immediately on ice packs in a cooler (4 °C) to preserve the physiological and biochemical state of the cells at the time of collection.
6. Fix the remaining samples with Lugol's Iodine (e.g., for ecological counts) to a final concentration of 1%.
7. Rinse the water sampler with distilled water before collecting the next sample.

3.3 Plankton Net Collection Procedure

1. Attach the cod-end (collecting vial) firmly to the plankton net's end.
2. Throw and recover the net, five to ten times in a steady hand over hand motion from shore or tow it from a boat at a slow, steady speed for two to five minutes at a depth of 0.2 to 0.5 meters.
3. Lift the net out of the water without allowing the cod-end submerge.
4. Rinse the inside of the net from the outside with distilled water to remove adhering plankton from the cod-end.
5. Detach the cod-end, if the sample volume is large, let it settle for 30–60 minutes and carefully remove the supernatant to concentrate the sample.
6. Gently transfer the concentrated sample into a prelabeled container and place it immediately in the cooler on ice.
7. Rinse the net and cod-end thoroughly with distilled water to prevent cross-contamination.
8. Transport all samples to the laboratory in the cooler on ice packs as quickly as possible.
9. Process samples immediately upon return for microscopic examination or inoculation into culture media.

3.4 Observation

Both direct water samples and plankton net samples were studied for microscopy characteristics during the observation period. The direct water sample appeared clean or faintly green, indicating a low level of suspended algal biomass. In contrast, the plankton net sample captured a concentrated solution that ranged from dark green to brownish, indicating a higher density of microalgae and other planktonic organisms. A drop of each container of mixed sample was placed on a sterile microscope slide, covered by a coverslip, and viewed with a light microscope at magnifications of 10×, 40×, and 100×. *Chlorella* species were detected under a microscope as spherical, bright green, nonmotile cells with a diameter of about 2–10 μm. Each cell's chloroplast was obviously cup-shaped, with a definite spherical pyrenoid visible at higher magnification. Analysis showed that the plankton net sample had a high concentration of *Chlorella* cells, with 10 to 50 cells per field of view at 100× magnification. At 100× magnification, the direct water sample often included only 0–5 *Chlorella* cells per field, indicating a much lower density. Furthermore, the plankton net sample contained a more diversified array of other planktonic forms, including rotifers, diatoms, cyanobacteria, and other green algae, illustrating how well the microalgal population was focused.

4 Notes

1. For optimal *Chlorella* cell recovery in natural freshwater habitats, use both direct water sampling and plankton net collection, especially when populations are low.
2. Important sampling data to capture include water temperature, pH, depth, GPS coordinates, turbidity, and collection time. This data explains the abundance of *Chlorella*.
3. Each sample container should have the date, site code, depth, and collector initials clearly marked. To keep an eye on sample handling and storage, keep an accurate list.
4. If microscopic analysis is postponed, samples should be preserved right away using formalin or Lugol's iodine. This allows *Chlorella* cells to keep their shape.
5. Concentrated samples can be rapidly filtered or centrifuged to provide pure biomass for developing cultures. In a lab setting, inoculate it into sterile microalgal growth media [7].

References

1. Kessler E (1976) Comparative physiology, biochemistry, and the taxonomy of Chlorella (Chlorophyceae). Plant Syst Evol 125(3): 129–138
2. Field CB, Behrenfeld MJ, Randerson JT, Falkowski P (1998) Primary production of the biosphere: integrating terrestrial and oceanic components. Science 281(5374):237–240
3. Khan MI, Shin JH, Kim JD (2018) The promising future of microalgae: current status, challenges, and optimization of a sustainable and renewable industry for biofuels, feed, and other products. Microb Cell Factories 17(1):36
4. Andersen RA (ed) (2005) Algal culturing techniques. Elsevier Academic Press

5. Palanivel P, Rajendran PJ, Thajuddin F, Elumalai A, Dharumadurai D, Muralitharan G (2026) Correlating the water quality parameters on algal diversity through multivariate statistical analysis and identifying the bioindicator of water pollution in the freshwater ecosystems of Dakshina, Karnataka. Water Air Soil Pollut 237(3): 168
6. Wetzel RG (2001) Limnology: lake and river ecosystems, 3rd edn. Academic Press
7. Barsanti L, Gualtieri P (2014) Algae: anatomy, biochemistry, and biotechnology (2nd ed). CRC Press

Chapter 2

Isolation of *Chlorella* spp. from Natural Water Using Serial Dilution and Micropipette Techniques

Reehana Nazar, Mohamed Imran MohamedYousuff, Thajuddin Nooruddin, and Dhanasekaran Dharumadurai

Abstract

Chlorella spp. is a unicellular green microalga known for its fast growth, simple structure, and high photosynthetic activity. In this study, *Chlorella* was collected from a freshwater source and purified using serial dilution followed by single-cell selection with a micropipette. Serial dilution helped reduce the number of associated microorganisms, while the micropipette method enabled the careful picking of individual algal cells. Together, these steps provided a dependable procedure for obtaining pure *Chlorella* cultures suitable for laboratory and biotechnological work. All media and glassware were sterilized before use, and UV light was used to maintain aseptic conditions. The cultures were incubated at 25–28 °C under controlled light (2000–4000 lux, 12:12 light–dark cycle). Green colonies were visible within 4–5 days and reached nearly 5 mm in diameter after 8–10 days. The purified colonies were transferred to liquid medium and examined microscopically to confirm axenic growth.

Key words *Chlorella* spp., Isolation, Serial dilution, Micropipette technique, Axenic culture

1 Introduction

Microalgae like *Chlorella* spp. are widely recognized for their biotechnological potential because of their uncomplicated cellular organization and the range of useful biochemical compounds they produce. As members of the division Chlorophyta, they contribute significantly to several applied fields, including biofuel production, treatment of wastewater, and the formulation of nutraceuticals and other environmentally friendly products. Obtaining pure cultures from natural water bodies, however, is often difficult, as these samples usually contain a mixture of many microorganisms [1].

To overcome this, different purification approaches have been developed, including serial dilution, micropipette-based cell picking, and plate streaking [2]. Serial dilution is straightforward and helps reduce unwanted microbial populations, but pure colonies

Dhanasekaran Dharumadurai (ed.), *Chlorella Genomics, Cultivation, and Bioenterprise: Green Economy through Algal Innovations*, Springer Protocols Handbooks, https://doi.org/10.1007/978-1-0716-5276-3_2,
© The Author(s), under exclusive license to Springer Science+Business Media, LLC, part of Springer Nature 2026

are not always guaranteed because contaminants may reappear during longer incubation. Micropipette isolation, in contrast, makes it possible to select individual algal cells with accuracy, though it demands careful handling and is more time-consuming. Since each method has its own drawbacks, using both together can greatly improve the chances of obtaining a clean algal culture.

In the present work, *Chlorella* spp. were collected from a freshwater source and first subjected to serial dilution to lower the level of associated microbes. Single cells were then isolated using a micropipette, and the plates were incubated under suitable conditions to encourage algal growth [3]. Distinct green colonies appeared within a few days, after which they were further purified and examined microscopically to confirm the absence of bacterial and fungal contaminants.

This combined approach, serial dilution followed by careful single-cell isolation proved to be a practical and dependable way to obtain clean cultures of *Chlorella* spp. [4]. The method provides axenic cultures that can be used for physiological and biochemical studies, molecular work, and various industrial applications in algal biotechnology.

2 Materials

2.1 Equipment

1. Sterile sampling bottles (500 mL).
2. Laminar flow hood.
3. Autoclave.
4. Light incubator (capable of 2000–4000 lux; 25–28 °C, 12: 12 h light: dark cycle).
5. Inverted microscope (for micropipette isolation).
6. Compound light microscope.
7. Sterile micropipette tip/a sterile inoculating loop or a sterile toothpick.
8. Sterile spreaders.
9. Sterile Petri dishes, Erlenmeyer flasks, test tubes.

2.2 Reagents/Media

1. Chu 10 Medium.
2. Agar.
3. Sterile distilled water.
4. 70% Ethanol for surface sterilization.
5. Sterile saline (0.85% NaCl) or sterile Chu 10 for dilutions.

Chu 10 Medium Composition

Ingredients	Milligrams/liter (mg/L)
Calcium nitrate	40
Magnesium sulphate	25
Dipotassium hydrogen phosphate	5
Sodium carbonate	20
Sodium silicate	25
Iron (II) chloride	8
Trace metal solution	1

Formula adjusted, standardized to suit performance parameters

Trace Metal Solution

Substance	Final concentration in 1 L medium (mg)
Boric acid	2.86
Manganese chloride	1.81
Zinc sulphate	0.222
Sodium molybdate	0.390
Copper sulphate	0.079
Cobalt nitrate	0.0494

3 Methods

1. Collect freshwater surface samples in sterile bottles, keep them cool, and process within 4–6 hours. A small drop of the sample is examined under the microscope to confirm the presence of *Chlorella* cells.
2. Prepare Chu 10 Medium agar and sterilize all glassware and media by autoclaving at 121 °C for 15–20 minutes. The molten medium is then poured into sterile Petri plates and allowed to solidify.
3. Label sterile test tubes with dilution factors 10^{-1} to 10^{-5} and add 9 mL of sterile distilled water to each tube.
4. Transfer 1 mL of the original water sample into the 10^{-1} tube using a sterile micropipette tip and mix thoroughly.
5. Continue serial dilutions by transferring 1 mL from each previous dilution to the next tube (up to 10^{-5}), ensuring proper mixing at every step.

6. Aliquots of 100–200 μL from the 10^{-4} and 10^{-5} dilutions are spread onto separate Chu 10 agar plates.
7. Sterilize a glass spreader by flaming, cool it, and distribute the inoculum evenly over the agar surface.
8. Incubate the plates in an inverted position at 25–30 °C under 2000–4000 lux light with a 12:12 h light–dark cycle for 2–4 weeks.
9. After incubation, observe the plates for distinct green colonies and examine them under a microscope to confirm *Chlorella* morphology.
10. A droplet of culture from a selected colony is placed on a sterile concave slide.
11. Under a microscope, use a fine-tipped micropipette to pick and isolate a single *Chlorella* cell.
12. Transfer the isolated cell into a fresh droplet of sterile Chu 10 medium on another slide and repeat the washing process several times to remove contaminants.
13. After washing, transfer the single cell into a sterile tube containing liquid Chu 10 medium or onto a new Chu 10 medium agar plate.
14. Incubate the culture under the same conditions as before until visible growth appears.
15. Examine the culture microscopically to confirm axenic (bacteria-free) growth. Repeat micropipette isolation if necessary to ensure purity (Fig. 1).

3.1 Observations

Green *Chlorella* colonies appear on Chu 10 agar plates within 4–5 days of incubation, with clearer and less contaminated growth in the 10^{-4} and 10^{-5} dilutions. The colonies were circular, smooth and reached about 4.8–5.2 mm in diameter by the eighth to tenth day. Microscopic observation reveals typical *Chlorella* cells—small,

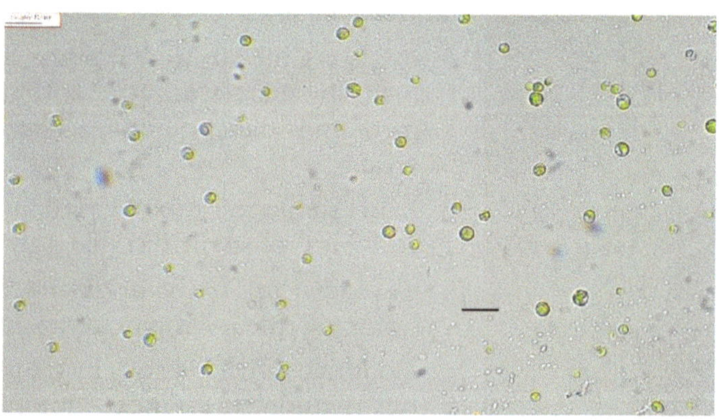

Fig. 1 Microscopic view of purified spherical *Chlorella* cells

spherical and bright green. The serial dilution process effectively reduces bacterial and fungal contamination, while the micropipette technique enables accurate isolation of individual algal cells, producing clean, pure cultures. The subcultured cells in liquid Chu 10 medium form a uniform green suspension without visible contaminants.

4 Notes

1. Maintaining strict aseptic conditions is essential to prevent bacterial overgrowth [1].
2. Performing multiple washing steps during micropipette isolation greatly improves the chances of obtaining axenic cultures [2].
3. Serial dilution alone does not ensure purity; it must always be followed by microscopic verification [3].
4. Light intensity plays an important role in determining colony color and growth rate [4].
5. Chu 10 medium provides optimal nutritional support for freshwater green microalgae [2, 3].

Acknowledgments

The author, Reehana Nazar, acknowledges the Secretary and Correspondent, the Principal and PG & Research Department of Microbiology, Jamal Mohamed College (Autonomous), Tiruchirappalli, Tamil Nadu, India, for their inherent support in preparing this data.

References

1. Fernandez-Valenzuela S, Chávez-Ruvalcaba F, Beltran-Rocha J, San Claudio P, Reyna-Martínez R (2021) Isolation and culturing axenic microalgae: mini–review. Open Microbiol J 15: 111
2. Andersen RA (2005) Algal culturing techniques. Elsevier Academic Press, New York, pp 92–93
3. Lee K, Eisterhold ML, Rindi F, Palanisami S, Nam PK (2014) Isolation and screening of microalgae from natural habitats in the midwestern United States of America for biomass and biodiesel sources. J Nat Sci Biol Med 5(2): 333–339
4. Haque Z, Desai ND (2024) Isolation of *Chlorella vulgaris* from water sample of Bhagva river, India. JETIR 11(5):693–697

Chapter 3

Purification of Axenic *Chlorella* spp. Cultures Using Streak Plate and Antibiotic Treatment

Reehana Nazar, Mohamed Imran MohamedYousuff, Thajuddin Nooruddin, and Dhanasekaran Dharumadurai

Abstract

Chlorella spp., a unicellular green microalga, holds significant biotechnological importance owing to its rapid growth, rich nutritional profile, and ability to thrive under diverse environmental conditions. The present study aims to purify *Chlorella* spp. isolated from freshwater sources using two complementary methods—the streak plate technique and antibiotic treatment. However, bacterial and fungal contaminants frequently interfere with its large-scale cultivation and hinder accurate physiological and genetic analyses. The streak plate method on modified solid media enables the isolation of discrete algal colonies, while antibiotic treatment efficiently suppresses bacterial contamination without affecting algal growth. Optimization of culture conditions such as temperature, light intensity, and organic nutrient supplementation further enhanced colony formation and minimized contamination. This integrated approach offers a reliable, efficient, and rapid method for obtaining axenic *Chlorella* cultures, facilitating their use in biofuel production, nutraceutical formulation, and environmental biotechnology applications.

Key words *Chlorella* spp., Purification, Streak plate method, Axenic culture, Antibiotic treatment

1 Introduction

Microalgae, particularly *Chlorella* species, are gaining global importance due to their vast applications in food, feed, pharmaceuticals, and biofuel industries. *Chlorella*, a unicellular green alga belonging to the division *Chlorophyta*, is well-known for its high photosynthetic efficiency and rich biochemical composition, including proteins, lipids, pigments, and vitamins. Maintaining pure (axenic) cultures is essential for the successful cultivation and utilization of *Chlorella*, as bacterial and fungal contaminants can interfere with algal metabolism, nutrient uptake, and growth [1]. Environmental isolates often coexist with diverse microbial populations, making purification a challenging, yet vital step for accurate physiological and molecular studies, as well as for industrial applications.

Dhanasekaran Dharumadurai (ed.), *Chlorella Genomics, Cultivation, and Bioenterprise: Green Economy through Algal Innovations*, Springer Protocols Handbooks, https://doi.org/10.1007/978-1-0716-5276-3_3,
© The Author(s), under exclusive license to Springer Science+Business Media, LLC, part of Springer Nature 2026

Various purification techniques have been developed, and among these, the streak plate method is widely used due to its simplicity and efficiency in obtaining discrete algal colonies, especially when combined with selective media and antibiotics [2]. In addition to mechanical isolation, antibiotic treatment serves as an effective complementary strategy for suppressing bacterial contaminants without affecting algal growth. Mixtures of antibiotics with different modes of action to target different organisms are commonly used as well [3, 4]. The combined use of streak plating and antibiotic treatment provides a synergistic method for rapid and reliable purification of *Chlorella* cultures. In this study, *Chlorella* spp. isolated from a freshwater habitat were purified using streak plate and antibiotic treatment methods under controlled conditions (25–28 °C temperature, 2000–4000 lux light intensity, and a 12:12 light–dark cycle). Pure colonies were subcultured and microscopically examined to confirm the absence of contaminants.

2 Materials

1. Chu 10 Medium agar plates.
2. Sterile distilled water.
3. Antibiotics—Ampicillin, Chloramphenicol, Gentamycin, Kanamycin, and Tetracycline.
4. 0.22-μm syringe filters.
5. Trace metal salts: Zinc chloride, Cobalt chloride, Ammonium molybdate tetrahydrate, Copper sulfate, Hydrochloric acid.
6. Ethylenediaminetetraacetic acid (EDTA)—chelating agent.
7. Inoculating loop.
8. Bunsen burner.
9. Laminar airflow cabinet.
10. Sterile flasks.
11. Centrifuge.
12. Micropipettes and sterile tips.
13. Nutrient Agar (NA) plates.
14. Compound microscope.
15. Culture room (20–28 °C, 12:12 h light–dark photoperiod).

3 Methods

3.1 Streak Plate Method

1. Place *Chlorella* culture bottles inside a laminar flow cabinet for 30 minutes to allow cells to settle.

2. Sterilize the inoculating loop by flaming until red hot; cool for about 1 minute.
3. Collect a loopful of algal culture from the bottom of the bottle.
4. Gently streak one-fourth of the agar plate surface in parallel lines.
5. Re-flame and cool the loop; streak from the edge of the first quadrant to the second quadrant in a zigzag pattern.
6. Repeat for the third and fourth quadrants without overlapping previous streaks.
7. Incubate plates at 25 ± 2 °C under 2000–4000 lux light intensity for 7–20 days.
8. Observe plates for isolated green colonies in the final quadrants.
9. Pick a single colony using a sterile loop and examine microscopically for purity.
10. Transfer pure colonies into sterile liquid Chu 10 Medium for growth and enrichment.
11. Repeat streaking if mixed growth is observed until a unialgal culture is achieved (Fig. 1).

3.2 Antibiotic Treatment for Axenization

1. Prepare antibiotic stock solutions by dissolving lyophilized powders in sterile distilled water or ethanol (based on solubility).
2. Filter-sterilize the antibiotic solutions using a 0.22-μm syringe filter.

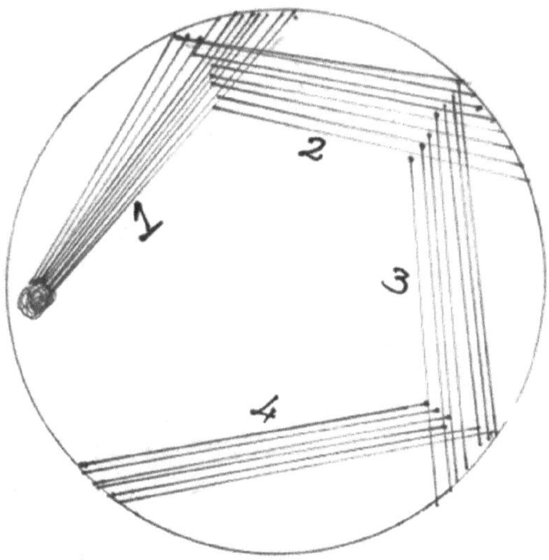

Fig. 1 Streak plate method

Table 1
Frequently used antibiotics for elimination of bacteria in microalgal cultures [5]

Antibiotic	Recommended working concentration (μg/mL)
Ampicillin	50–300
Chloramphenicol	10–100
Gentamycin	10–100
Kanamycin	10–150
Tetracycline	10–100

3. Wash algal cells with sterile distilled water and centrifuge at 3500 rpm for 15 minutes.
4. Resuspend algal pellets in sterile medium.
5. Add antibiotics at recommended working concentrations as shown in Table 1.
6. To avoid harmful effects of antibiotics on algal growth, test different concentrations before final use.
7. Treat algal cultures with selected antibiotic concentrations (0.5–10%) and incubate for 3–5 days under 20–28 °C with a 12:12 h light–dark cycle.
8. Observe daily for algal growth, color, and morphology under the microscope.
9. Plate 1 mL of treated culture on NA plates to confirm absence of bacteria.
10. Incubate plates in the dark at 25 °C for 48–72 hours.
11. Select cultures showing no contamination and healthy algal cells.
12. Maintain purified *Chlorella* as axenic stock for future use.

3.3 Observation

After 7–10 days of incubation, distinct green colonies of *Chlorella* were observed on Chu 10 Medium agar plates, especially in the third and fourth streak quadrants. Microscopic examination confirmed the characteristic spherical, bright green morphology of *Chlorella* cells. Antibiotic treatment with gentamycin (50 μg/mL) and kanamycin (100 μg/mL) effectively eliminated bacterial contaminants without affecting algal growth. Cultures grown in antibiotic-treated media appeared uniformly green, showing no bacterial growth on NA plates. The purified colonies were successfully subcultured into liquid Chu 10 Medium to obtain axenic *Chlorella* cultures.

4 Notes

1. Repeated streaking helps in securing a well-isolated unialgal colony [2].
2. Antibiotic concentrations should be adjusted according to the contamination load [1, 4].
3. Excessive antibiotic use can affect chlorophyll production, so only the lowest effective dose is advisable [3].
4. All procedures must be carried out aseptically inside a laminar airflow cabinet [2].

Acknowledgments

The author, Reehana Nazar, acknowledges the Secretary and Correspondent, the Principal and PG & Research Department of Microbiology, Jamal Mohamed College (Autonomous), Tiruchirappalli, Tamil Nadu, India, for their inherent support in preparing this data.

References

1. Droop MR (1967) A procedure for routine purification of algal cultures with antibiotics. Br Phycol Bull 3(2):295–297
2. Guillard RRL (2005) Purification methods for microalgae. In: Andersen RA (ed) Algal culturing techniques. Elsevier Academic Press, pp 117–132
3. Gonzalez-Pleiter M, Gonzalo S, Rodea-Palomares L, Leganes F, Rosal R, Boltes K, Marcos E, Fernandez-Piñas F (2013) Toxicity of five antibiotics and their mixtures towards photosynthetic aquatic organisms: implications for environmental risk assessment. Water Res 47: 2050–2064
4. Molina-Cardenas CA, Sanchez-Saavedra MDP, Licea-Navarro AF (2016) Decreasing of bacterial content in Isochrysis galbana cultures by using some antibiotics. Rev Biol Mar Oceanogr 51(1):101–112
5. Ermakova-Gerdes S, Vermaas W (1999) Development of a psbA-/psbD- strain of Synechocystis sp. PCC 6803 for simultaneous mutagenesis of the D1 and D2 proteins of photosystem II. In: The phototrophic prokaryotes. Kluwer Academic Publishers, Dordrecht, pp 1–60

Part II

Taxonomic Characterization and Identification of *Chlorella* Species

Chapter 4

Wet Mount Preparation and Microscopic Examination of *Chlorella vulgaris*, *Chlorella sorokiniana*, *Chlorella pyrenoidosa*, *Chlorella minutissima*, *Chlorella protothecoides*

Priyanka Jayam Rajendran, Bhuvaneshwari Srinivasan, Gowtham Neethimohan, and Dhanasekaran Dharumadurai

Abstract

Chlorella species are identified using wet mount preparation and light microscopy. Wet mount techniques allow for the visualization of living algal cells while they are hydrated. This approach preserves their inherent structure without distortion due to fixing and staining. The produced slides are examined under a compound microscope, and their morphology is compared to conventional algae taxonomic references. Shape, size, chloroplast structure, and reproductive habits are all used to identify organisms. *Chlorella* cells studied are typically spherical or subspherical, measuring between 2 and 10 μm. A large cup-shaped chloroplast with a visible pyrenoid is present, confirming *Chlorella* common features. The cells appeared solitary, lacked mucilage, and reproduced asexually via autospores. The wet mount technique enabled clear visualization of live cellular characteristics, which are critical for differentiating *Chlorella* species and establishing their morphological identification.

Key words Wet mount, *Chlorella* sp., Cell shape, Size

1 Introduction

The genus *Chlorella* is a group of unicellular green microalgae belonging to the division *Chlorophyta* [1]. They are commonly found in freshwater, soil, and wastewater environments. *Chlorella* cells can be spherical or ellipsoidal, with diameters ranging from 2 to 15 μm. *Chlorella* has no sexual life cycle and reproduces exclusively through asexual autospore formation. When autospores mature, the mother cell wall ruptures, releasing them simultaneously. The number of autospores produced from a single mother cell might range from two to sixteen. Typically, *Chlorella* contains a cup-shaped chloroplast situated peripherally in the cytoplasm. The

Dhanasekaran Dharumadurai (ed.), *Chlorella Genomics, Cultivation, and Bioenterprise: Green Economy through Algal Innovations*, Springer Protocols Handbooks, https://doi.org/10.1007/978-1-0716-5276-3_4,
© The Author(s), under exclusive license to Springer Science+Business Media, LLC, part of Springer Nature 2026

nucleus is found near the cytoplasmic membrane, and the mitochondria are tightly linked with the chloroplast [7]. The *wet mount preparation* technique is used to observe live algal cells in their natural, hydrated state without fixation or staining, enabling visualization of size, shape, chloroplast structure, and cell arrangement. Within the newly analyzed strains, seven species are observed with a surrounding mucilaginous envelope. The colonial life form is often lost in culture, disintegrating into single cells [2].

2 Materials

1. Clean glass slides.
2. Cover slips.
3. Droppers or micropipettes.
4. Algal culture sample.
5. Compound microscope.
6. Tissue paper.
7. Immersion oil (for 100× observation).

3 Methods

1. Use 70% ethanol to clean a cover slip and a glass slide.
2. Place a small drop (5–10 μL) of the algal suspension at the center of the slide.
3. Then place one edge of a cover slip onto the slide near the drop.
4. Allowing the cover slip to fall onto the drop, enclosing the sample between the slide and the cover slip [6].
5. The wet mount is now ready to examine (Fig. 1).
   (a) Then observe this prepared slide under a compound microscope at 40× magnification [4].

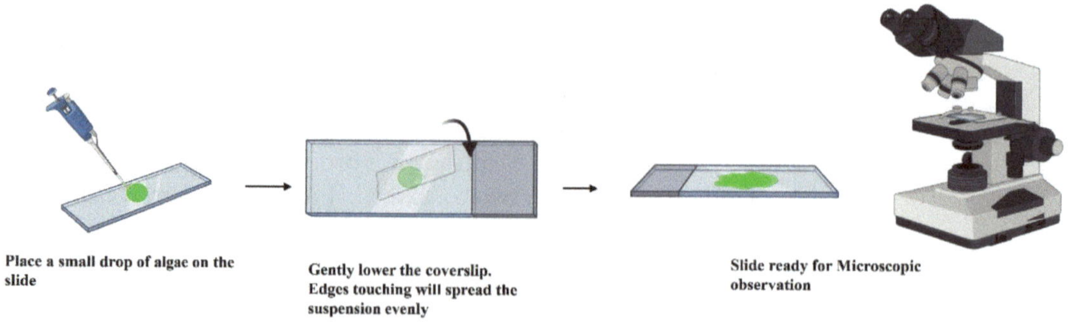

Fig. 1 Wet mount slide preparation for microalgae examination

Table 1
Taxonomic descriptions and morphological features of key *Chlorella* strains

Morphological feature	Cell shape	Chloroplast morphology	Presence of pyrenoid	Cell wall	Reference
Chlorella vulgaris	Spherical microscopic cell with 2–10 μm	Single, cup-shaped	Absent or inconspicuous	Smooth, thin Lack of flagellum	[11]
Chlorella sorokiniana	Spherical with 2–4.5 μm diameter	Single, parietal (plate-like)	Absent	Smooth, thin	[8]
Chlorella pyrenoidosa	Cells spherical, 3–5 μm diameter	Single, parietal mantle shaped	The pyrenoid is conspicuous and single	Smooth, thin	[3]
Chlorella minutissima	Spherical with 2 to 3·8 f.tm in diameter	Single, cup-shaped chloroplast	The pyrenoid is often absent or indistinct. Spherical pyrenoid	Smooth, trilaminar outer layer	[5]
Chlorella protothecoides or Auxenochlorella protothecoides	Spherical to oval with 1.5–12.0 μm	Cup-shaped,	The pyrenoid is typically absent.	Smooth, thick	[12]

(b) Categorize the microalgae in the water samples by their shape and size (Table 1). Use traditional taxonomic books and existing research for identification [3, 10] (Fig. 2).

3.1 Observation

1. The morphology is recorded using a photomicroscopic device (MCX 500, Micros, Austria) to take photomicrographs [9].
2. According to morphological features, *Chlorella* species are found to have spherical to subspherical cells with diameters ranging from 2 to 10 micrometers.
3. A distinguishing feature of the genus is that each cell has one cup-shaped chloroplast, each with a clear pyrenoid. The cells are solitary, lack a mucilaginous covering, and reproduce asexually by autospores [11].

4 Notes

Slides and coverslips must be properly cleaned to avoid artifacts and provide good visualization. Morphological identification should preferably be combined with molecular markers, such as 18S

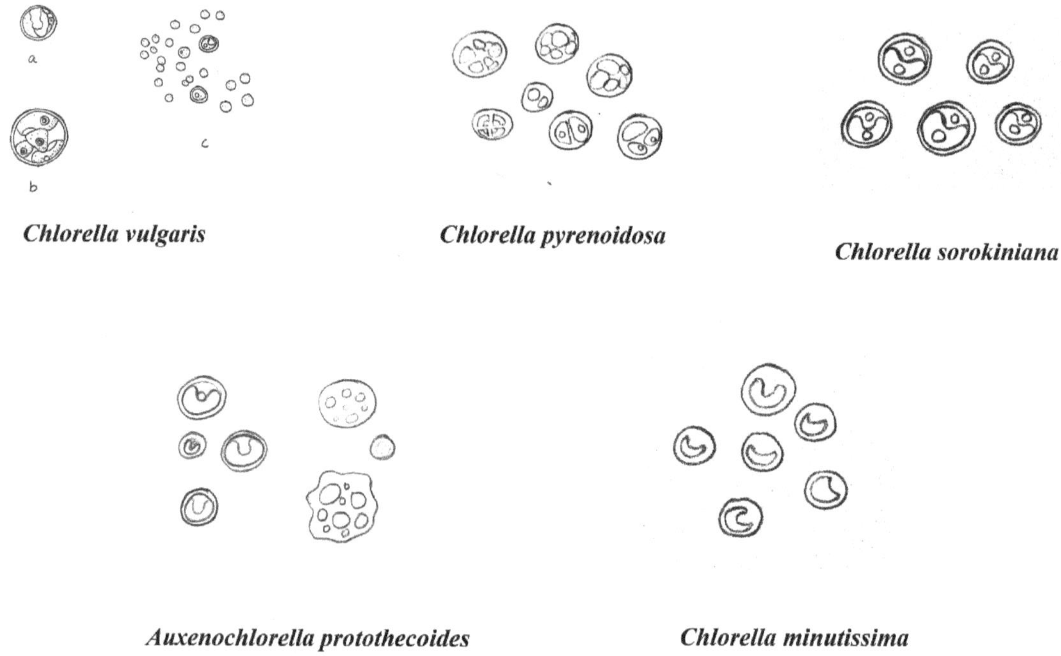

Fig. 2 Morphological observation of *Chlorella* sp.

rRNA, to confirm species identification. Placing the coverslip at an angle helps to avoid air bubbles, which can disrupt microscopic observation.

Data Availability Statement All data generated or analyzed during this study are provided in the chapter and its supplementary information files.

Ethics Approval Not applicable.

Competing Interests The authors declare that they have no competing financial or nonfinancial interests that could have influenced the work reported in this chapter.

Declaration of Generative AI and AI-Assisted Technologies in the Writing Process During the preparation of this work, the author(s) used AI to improve the readability of the chapter. After using this tool, the author(s) reviewed and edited the content as needed and take(s) full responsibility for the content of the publication.

Funding

Department of Biotechnology (Govt. of India) for the establishment of the National Repository for Microalgae and Cyanobacteria—Freshwater & Marine (NRMC F & M) (Phase II) (BT/ PR29901/PBD26/694/2018) research. Chief Minister Research Grant (CMRG) for the financial support of research by the Government of Tamil Nadu, India, for the financial support (CMRG 2023–2024- Project No. CMRG2023BMBO7047).

References

1. Baldisserotto C, Demaria S, Arcidiacono M, Benà E, Giacò P, Marchesini R, Pancaldi S (2023) Enhancing urban wastewater treatment through isolated *Chlorella* strain-based phytoremediation in centrate stream: an analysis of algae morpho-physiology and nutrients removal efficiency. Plants 12(5):1027
2. Bock C, Krienitz L, Pröschold T (2011) Taxonomic reassessment of the genus *Chlorella* (Trebouxiophyceae) using molecular signatures (barcodes), including description of seven new species. Fottea 11(2):293–312
3. Chick H (1903) A study of a unicellular green alga, occurring in polluted water, with especial reference to its nitrogenous metabolism. Proc R Soc Lond 71(467–476):458–476
4. Clason T, Acker F, Morales E, Marr L (2002) Analysis of diatoms on microscope slides prepared from USGS NAWQA program algae samples. Protocols for the analysis of algal samples collected as part of the US Geological Survey National Water-Quality Assessment program, 61
5. Dempsey GP, Lawrence D, Cassie V (1980) The ultrastructure of Chlorella minutissima Fott et Nováková (Chlorophyceae, Chlorococcales). Phycologia 19(1):13–19
6. Francoeur SN, Rier ST, Whorley SB (2013) Methods for sampling and analyzing wetland algae. In: Wetland techniques: volume 2: organisms. Springer Netherlands, Dordrecht, pp 1–58
7. Liu J, Chen F (2014) Biology and industrial applications of Chlorella: advances and prospects. In: Microalgae biotechnology, pp 1–35
8. Lizzul AM, Lekuona-Amundarain A, Purton S, Campos LC (2018) Characterization of Chlorella sorokiniana, UTEX 1230. Biology 7(2): 25
9. Narchonai G, Arutselvan C, LewisOscar F, Thajuddin N (2019) Deciphering the microalgal diversity and water quality assessment of two urban temple ponds in Pondicherry, India. Biocatal Agric Biotechnol 22:101427
10. Prescott GW (1964) How to know the freshwater algae
11. Safi C, Zebib B, Merah O, Pontalier PY, Vaca-Garcia C (2014) Morphology, composition, production, processing and applications of Chlorella vulgaris: a review. Renew Sust Energ Rev 35:265–278
12. Zhao L, Dai J, Wu Q (2014) Autophagy-like processes are involved in lipid droplet degradation in Auxenochlorella protothecoides during the heterotrophy-autotrophy transition. Front Plant Sci 5:400

Chapter 5

Examination of Live and Dead *Chlorella* Cells Using Fluorescence Staining

Priyanka Jayam Rajendran, Prakash Palanivel, and Dhanasekaran Dharumadurai

Abstract

Microalgae viability is crucial for biotechnological and environmental applications. An effective dual-fluorescence identification technique for differentiating between living and dead *Chlorella* cells is described in this chapter. The method uses fluorescein diacetate (FDA), which can be hydrolyzed by esterases in living cells to produce green fluorescence. Propidium Iodide (PI), on the other hand, can enter dead cells or cells with damaged membranes and bind to nucleic acids, producing red fluorescence. Both live (untreated) and dead (heat-treated and ethanol-treated) cells are used for testing. Using an epifluorescence microscope, researchers look at the dyed cells to determine which populations are viable and which are not. This procedure provides a straightforward, efficient, and quantitative method for evaluating microalgal viability in fields such as algal biotechnology, bioremediation, and ecotoxicology.

Key words Fluorescence staining, Fluorescein diacetate, Propidium iodide, Living and dead cell assay, Bioimage

1 Introduction

Chlorella sp., a single-celled microalga, has an average diameter of 4–10 μm. It can resist temperatures ranging from 3 to 30 degrees Celsius and light levels of 6000 to 12,000 Lux. *Chlorella* sp. can be found in freshwater, saltwater, soil, ponds, hot springs, and as plankton in the Arctic and Antarctic. *Chlorella* sp. exhibits great photosynthetic efficiency. It requires carbon dioxide, water, light, and trace amounts of minerals to generate energy quickly [2]. Because of their distinct features and the availability of natural compounds and pigments, microalgae have enormous promise for bioremediation, biofuels, food additives, ecological bioindication, and medicine. In these applications, assessing microalgae viability is an important stage in the culture and treatment process [3].

Dhanasekaran Dharumadurai (ed.), *Chlorella Genomics, Cultivation, and Bioenterprise: Green Economy through Algal Innovations*, Springer Protocols Handbooks, https://doi.org/10.1007/978-1-0716-5276-3_5,
© The Author(s), under exclusive license to Springer Science+Business Media, LLC, part of Springer Nature 2026

Determining the viability of microalgae is a crucial stage in their production and treatment for a variety of applications. In aquatic ecology and ecotoxicology, it's also crucial to create a simple technique to evaluate microalgae viability. This involves assessing the efficacy of an algaecide and researching the detrimental effects of pollutants on microalgae. Measuring conventional indicators of algal activity, such as cell density, pheophytin concentration, and chlorophyll concentration, is the most effective technique to determine if microalgal cells are alive or dead. Additional techniques include fluorescence staining with fluorescein diacetate, carboxyfluorescein diacetate, and SYTOX green, as well as chemical staining with dyes including neutral red, erythrite, Evans blue, trypan blue, and methylene blue [1]. In general, these techniques fall into one of two categories: (a) use a microscope to identify living from dead microalgae and (b) use a spectrophotometer or flow cytometry to get information on the survival rate of a large number of microalgae.

2 Materials

1. *Chlorella* sp. culture (log-phase, healthy cells).
2. BG11 medium ($NaNO_3$, K_2HPO_4, $MgSO_4 \cdot 7H_2O$, $CaCl_2 \cdot 2H_2O$, citric acid, Ferric ammonium citrate, Na_2CO_3, Trace Metal Solution:H_3BO_3, $MnCl_2 \cdot 4H_2O$, $ZnSO_4 \cdot 7H_2O$, $Na_2MoO_4 \cdot 2H_2O$, $CuSO_4 \cdot 5H_2O$, $Co(NO_3)_2 \cdot 6H_2O$).
3. Phosphate-buffered saline (PBS), pH 7.2–7.4.
4. Fluorescein diacetate (FDA) 5 mg/mL.
5. Propidium iodide (PI).
6. Stock: 1 mg mL^{-1} in sterile water.
7. Sterile distilled water.
8. 1.5 mL microcentrifuge tubes.
9. Glass microscope slides and coverslips.
10. Micropipettes and sterile tips.
11. Centrifuge.
12. Epifluorescence microscope.
13. Dark incubation box or aluminium foil (to protect dyes from light).

3 Methods

1. Use propidium iodide (PI) and fluorescein diacetate (FDA) to distinguish between living and dead cells. Prepare the fluorescein diacetate main stock at 5 mg/mL in DMSO. Store it at

−20 °C. Make the PI primary stock at 3 mg/mL in water. Store it at 4 degrees Celsius [5].

2. Use FDA and PI dual staining to distinguish between living and dead cells in a single sample.
3. Divide 50 ml *Chlorella* culture into two equal portions and centrifuge at 2450 × g for 1 minute.
4. Resuspend the pellets in 0.5 mL of culture media and 0.5 mL of pure ethanol.
5. After 30 seconds, alcohol-treated cells are centrifuged at 2450 × g for 1 minute and resuspended in 0.5 mL culture media.
6. Boil the cells for 15 minutes to apply heat therapy.
7. Ethanol-treated *Chlorella* cells or heat-treated *Chlorella* cells should be viewed as dead controls, whereas untreated cells should be treated as alive.
8. Cell suspensions can be combined to make mixtures containing 0, 25, 50, 75, or 100% live cells [4].
9. Add the FDA stock solution to the prepared *Chlorella* suspension in a ratio of 1:100 and incubate for 30 minutes at 22 °C.
10. Rinse the cell suspension twice with 1x NaHCO3 buffer before adding 25 μg mL − 1 of PI stock solution.
11. Incubate the suspension at 22 °C in the dark for 15 minutes [3].
12. Observe, measure, and record the staining outcome using a fluorescence microscope, cell counter, and flow cytometer (Fig. 1).

3.1 Observation

FDA-stained live *Chlorella* cells fluoresced bright green in the presence of blue light, indicating active intracellular esterase activity. Dead cells stained with PI fluoresced red due to penetration into cells with damaged membranes. Properly labelled cells showed no overlap of green and red signals. Mixed samples demonstrated both populations clearly. The cell viability percentage is measured by counting green (living) and red (dead) cells using a fluorescence microscope (Fig. 2).

4 Notes

Fluorescein diacetate (FDA) and propidium iodide (PI) are light-sensitive dyes. To avoid degeneration, always prepare and handle them in low light or using foil. To avoid contamination, handle microalgal cultures under clean conditions.

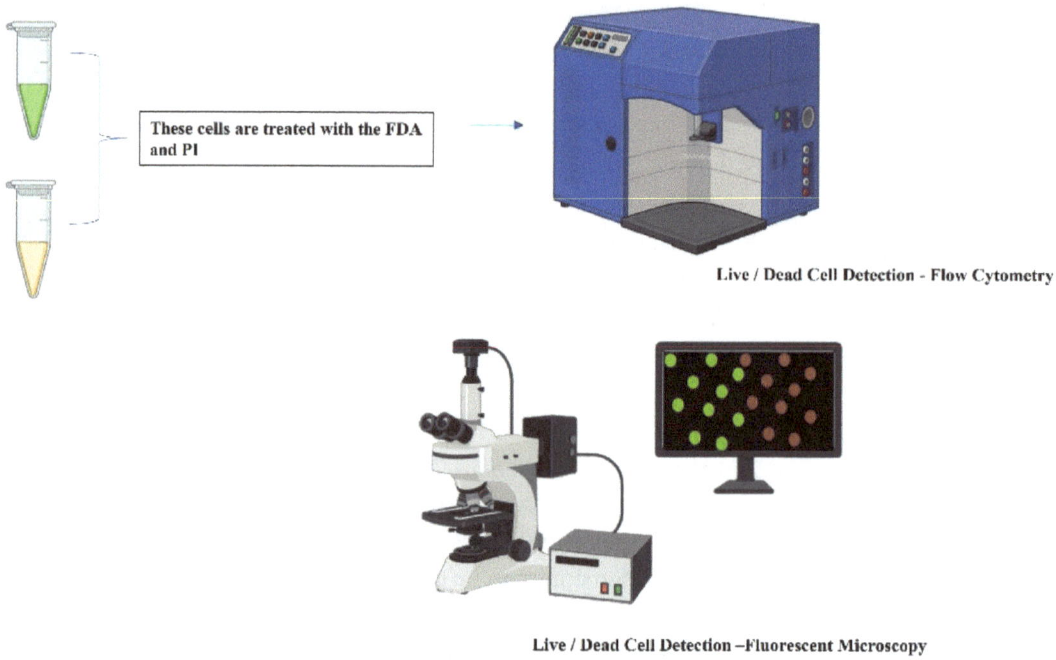

Fig. 1 Fluorescent and flow cytometric analysis of *Chlorella* viability

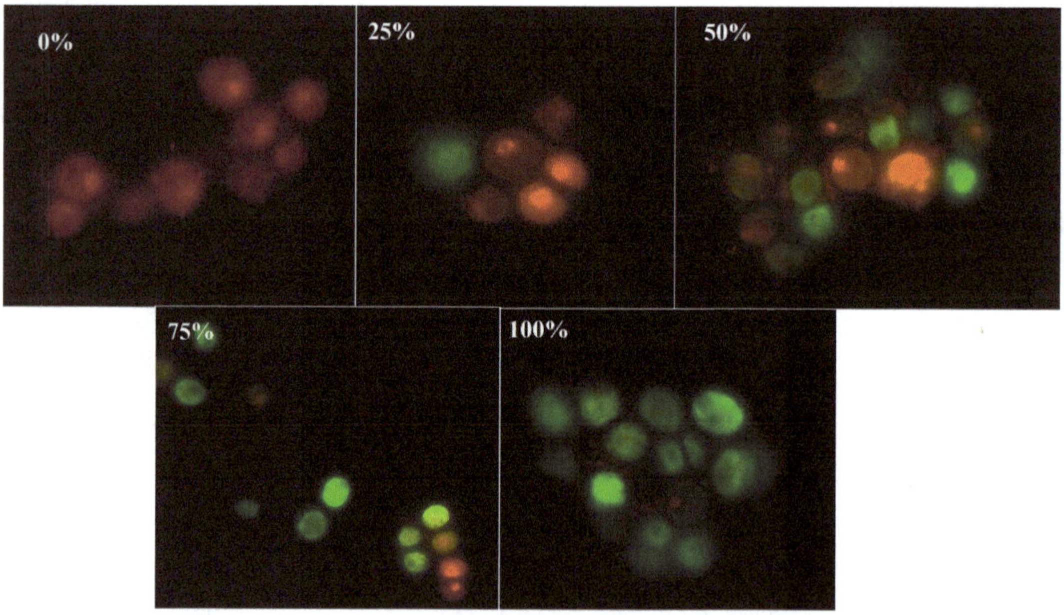

Fig. 2 Live and dead cell assay using FDA and PI with samples containing 0, 25, 50, 75, and 100% live cells

Data Availability Statement All data generated or analyzed during this study are provided in the chapter and its supplementary information files.

Ethics Approval Not applicable.

Author Contributions Conceptualization and Data collection: P. P., Writing original draft, data analyses and methodology: P.R., Supervision, Writing—review and editing: D.D. All contributors read the Manuscript and provided edits. All of the authors conceptualized the research. The article was read and approved by every author.

Competing Interests The authors declare that they have no competing financial or nonfinancial interests that could have influenced the work reported in this chapter.

Declaration of Generative AI and AI-Assisted Technologies in the Writing Process During the preparation of this work, the author(s) used AI to improve the readability of the chapter. After using this tool, the author(s) reviewed and edited the content as needed and take(s) full responsibility for the content of the publication.

Funding

Department of Biotechnology (Govt. of India) for the establishment of the National Repository for Microalgae and Cyanobacteria—Freshwater & Marine (NRMC F & M) (Phase II) (BT/ PR29901/PBD26/694/2018) research. Chief Minister Research Grant (CMRG) for the financial support of research by the Government of Tamil Nadu, India, for the financial support (CMRG 2023–2024- Project No. CMRG2023BMBO7047).

References

1. Dong D, Liu T, Liang D, Jin X, Qi Z, Li A, Ning Y (2021) Facile hydrothermal synthesis of *Chlorella*-derived environmentally friendly fluorescent carbon dots for differentiation of living and dead *Chlorella*. ACS Appl Bio Mater 4(4): 3697–3705
2. Lung WQC, Yeh HY, Yang SJ, Huang CY, Nan FH, Lee MC (2022) Delayed signs of UV-C damage to *Chlorella* sp. observed through fluorescent staining. Diversity 14(5):376
3. Reimann R, Zeng B, Jakopec M, Burdukiewicz M, Petrick I, Schierack P, Rödiger S (2020) Classification of dead and living microalgae *Chlorella vulgaris* by bioimage informatics and machine learning. Algal Res 48: 101908
4. Sato M, Murata Y, Mizusawa M, Iwahashi H, Oka SI (2004) A simple and rapid dual-fluorescence viability assay for microalgae. Microbiol Cult Coll 20(2):53–59
5. Yewalkar S, Wu T, Kuan D, Wang H, Li D, Johnson A et al (2019) Applicability of differential fluorescein diacetate and propidium iodide fluorescence staining for monitoring algal growth and viability. Waste Disposal Sustain Energy 1(3):199–206

Chapter 6

Identification of *Chlorella* and Bacterial Signaling Molecules

Priyanka Jayam Rajendran, Gowtham Neethimohan, Prakash Palanivel, and Dhanasekaran Dharumadurai

Abstract

The identification of *Chlorella* and associated bacterial signaling molecules offers insight into the interactions between algae and bacteria in microalgal growth systems. While microalgae are often grown as monocultures, they form dynamic microecosystems through nutrient exchange and biochemical communication. Quorum sensing (QS) has a significant impact on bacterial behavior and algal development. This chapter provides a procedure for detecting the key bacterial signaling molecules, N-acyl homoserine lactones (AHLs), and indole-3-acetic acid (IAA). N-acyl homoserine lactones (AHLs) production is assessed using biosensor-based chromogenic assays, thin-layer chromatography (TLC), and mass spectrometry. This method allows for the detection of the diversity and activity of signaling molecules. The chapter enhances our understanding of *Chlorella* bacterial communication and supports future uses in *Chlorella* biotechnology, bloom ecology, and microbial community engineering.

Key words Bloom ecology, Algae-bacteria association, Volatile signaling molecules, N-acyl homoserine lactones, Indole-3-acetic acid, Biosensor strain

1 Introduction

The microalgae culture system has never been composed of a single microalgae species, but rather of a unique, stable, and sophisticated microecological system that forms biologically during the microalgae growth process. Symbiosis, synergism, competition, and parasitism are all possible interactions between microalgae and bacteria (Fig. 1) [6]. Nutrient interaction, a fundamental type of algae–bacteria association, has been extensively studied for its ability to directly regulate algal and bacterial growth and metabolism [3]. However, there is insufficient research on the involvement of signal transduction via bacterial quorum sensing (QS) in the connection between algae and bacteria. Bacteria are able to detect the

Dhanasekaran Dharumadurai (ed.), *Chlorella Genomics, Cultivation, and Bioenterprise: Green Economy through Algal Innovations*, Springer Protocols Handbooks, https://doi.org/10.1007/978-1-0716-5276-3_6,
© The Author(s), under exclusive license to Springer Science+Business Media, LLC, part of Springer Nature 2026

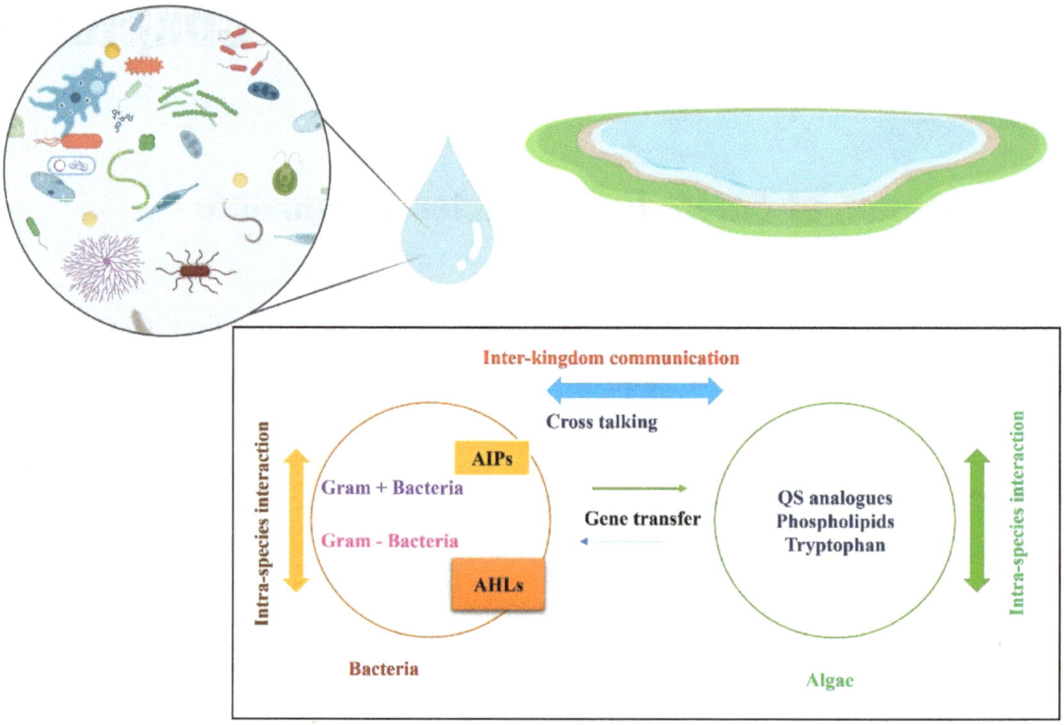

Fig. 1 *Chlorella* sp. and bacterial quorum sensing

presence of other bacteria by sensing changes in the concentration of signal molecules, and the expression of some related genes is regulated to provide varied biological functions [2]. N-acyl homoserine lactones (AHLs), which are found in a wide variety of Gram-negative bacteria, have been the most intensively studied bacterial quorum-sensing molecules. Meanwhile, signal exchange between bacteria and eukaryotes, including microalgae, has been demonstrated to be common [6]. Among the various forms of signaling molecules, indole-3-acetic acid (IAA) and N-acyl homoserine lactones (AHLs) are two scientifically proven signaling molecules from bacteria to microalgae that could be used as markers to estimate communication intensity. Indole-3-acetic acid (IAA) has been identified as a crucial signaling molecule between algae and bacteria. The exchange of the auxin indole-3-acetic acid (IAA) from bacteria increased microalgal cell division. Bacterial IAA modulated the algal cycling during signaling transmission. Similarly, AHLs play a key role in signaling between algae and bacteria [5]. Bacterial sensors are commonly employed to detect the presence of signaling molecules in the QS system. Although these biosensors cannot create pure signaling molecules, they can induce identifiable phenotypes like light emission, β-galactosidase activity, or pigment synthesis when exogenous active signaling molecules are added.

2 Materials

1. *Chlorella* culture (axenic or treated sample).
2. Bacterial isolates associated with *Chlorella*.
3. LB agar and LB broth.
4. Antibiotics (as required for biosensor strains).
5. QS biosensor bacterial strains (e.g., *Chromobacterium violaceum*, *Agrobacterium tumefaciens* A136, etc.).
6. Sterile distilled water.
7. X-gal (40 μg/mL).
8. LB medium with 0.8% agar for overlay plates.
9. TLC silica gel plates (60 F254).
10. Chloroform and methanol (analytical grade).
11. TLC developing chamber.
12. Liquid chromatography–mass spectrometry (LC–MS) system with C18 column.
13. Acetonitrile with 0.1% formic acid.
14. GC–MS instrument for volatile compound analysis.
15. Micropipettes, tips, conical tubes.
16. Incubator.
17. Petri dishes.
18. Vortex Mixer.
19. Centrifuge.

3 Methods

1. Examine the *Chlorella*-associated bacterial isolates for the production of AHLs using an agar plate diffusion assay in a biosensor system.
2. Grow active cultures of QS reporter bacterial strains in antibiotic containing LB agar and incubate the plates at 28 °C for 24 hours.
3. Take a loopful of overnight cultures from each plate and mix individually with 10 mL of sterile distilled water.
4. Prepare a thin plate using the biosensor strain's culture suspension for 2.5 mL and mix with 100 mL of cooled LB medium (0.8% agar) at 45 °C.
5. Add 40 μg/mL of X-gal as a visualizing agent in the LB medium.

6. Culture bacterial isolates inoculated in LB broth for 24 hours, then diluted them in sterile distilled water to an OD_{600} nm of 1.0.
7. Spot-inoculate approximately 5 μL of each test strain onto the prepared chromoplates and allow the spots to dry.
8. Incubate petri plates at 28 °C for 24 to 48 hours resulted in the synthesis of purple pigment violacein and blue coloration by β-galactosidase activity, indicating AHL generation [1].
9. Separate and visualize signaling molecules using TLC. Use a 60:40 chloroform–methanol solvent to generate extracts (5–20 μL) that are placed onto silica gel plates (60 F254).
10. Allow the plates to dry and cover the plates with soft agar containing biosensors. Record the color or fluorescence changes at various Rf positions to show the diversity of signaling molecules.
11. Perform chemical characterization using liquid chromatography and mass spectrometry. Load 10 μL of each sample onto a C18 reverse-phase column and elute with a water and acetonitrile mixture containing 0.1% formic acid.
12. Collect mass spectra in both positive and negative ion modes [4]. Conduct GC–MS analysis to detect volatile signaling molecules.

3.1 Observation

The biosensor tests generate clear visual responses to bacterial signaling chemicals. N-acyl homoserine lactone (AHL)-producing isolates produce purple pigment in *Chromobacterium violaceum, i*ndicating violacein synthesis, and blue coloration in β-galactosidase-based sensors, demonstrating QS activation. On TLC plates, separate colored or fluorescent spots develop at varying Rf values, indicating the presence of diverse signaling molecules. The LC-MS analysis reveals typical mass peaks that correspond to AHL and IAA derivatives, confirming their chemical identity. GC–MS profiles identify volatile QS-related chemicals produced by bacterial partners. Overall, these findings indicate that the bacteria found in *Chlorella* actively create signaling chemicals that can influence microalgal development and physiological responses.

4 Notes

TLC-biosensor overlay is a consistent method for visually mapping several signaling substances on one plate. Volatile compounds discovered by GC–MS may reveal other QS mechanisms beyond AHLs. Data should be examined in conjunction with algal physiological responses to have a complete knowledge of algae–bacteria interactions.

Data Availability Statement All data generated or analyzed during this study are provided in the chapter and its supplementary information files.

Ethics Approval Not applicable.

Author Contributions Conceptualization and Data collection: P. P., Writing original draft, Data analyses and methodology: P.R., Supervision, Writing—review and editing: D.D. All contributors read the Manuscript and provided edits. All of the authors conceptualized the research. The article was read and approved by every author.

Competing Interests The authors declare that they have no competing financial or nonfinancial interests that could have influenced the work reported in this chapter.

Declaration of Generative AI and AI-Assisted Technologies in the Writing Process During the preparation of this work, the author(s) used AI to improve the readability of the chapter. After using this tool, the author(s) reviewed and edited the content as needed and take(s) full responsibility for the content of the publication.

Funding

Department of Biotechnology (Govt. of India) for the establishment of the National Repository for Microalgae and Cyanobacteria—Freshwater & Marine (NRMC F & M) (Phase II) (BT/ PR29901/PBD26/694/2018) research. Chief Minister Research Grant (CMRG) for the financial support of research by the Government of Tamil Nadu, India, for the financial support (CMRG 2023–2024- Project No. CMRG2023BMBO7047).

References

1. Lade H, Paul D, Kweon JH (2014) Isolation and molecular characterization of biofouling bacteria and profiling of quorum sensing signal molecules from membrane bioreactor activated sludge. Int J Mol Sci 15(2):2255–2273
2. Li KY, Zhou JL, Vadiveloo A, Li JN, Zhou D, Wang JM et al (2025) Signal molecule-mediated algae-bacteria interactions and membrane-enabled separation synergize wastewater bioremediation. Chem Eng J 523:168368
3. Rajendran PJ, Nooruddin T, Dharumadurai D (2025) Metabarcoding of bacterial communities impacting the dynamics of cyanobacterial blooms in freshwater reservoirs, South India. Water Air Soil Pollut 236(1):66
4. Wang J, Quan C, Wang X, Zhao P, Fan S (2011) Extraction, purification and identification of bacterial signal molecules based on N-acyl homoserine lactones. Microb Biotechnol 4(4): 479–490

5. Wu X, Kong L, Feng Y, Zheng R, Zhou J, Sun J, Liu S (2024) Communication mediated interaction between bacteria and microalgae advances photogranulation. Sci Total Environ 914: 169975

6. Zhang B, Sun W, Su Y, Ren Q, Ji Z, Zhang A (2022) Utilization of N-Acyl homoserine lactone-secreting bacteria in algal environment to increase biomass accumulation of *Chlorella*. Bioenergy Res 15(4):2111–2121

Chapter 7

Isolation and Identification of Bacterial Symbionts from *Chlorella* Phycosphere Symbionts

Priyanka Jayam Rajendran and Dhanasekaran Dharumadurai

Abstract

The phycosphere is a dynamic environment that surrounds microalgal cells, is rich in organic exudates, and supports a diversity of bacterial species. *Chlorella* species release extracellular metabolites such as carbohydrates, amino acids, organic acids, and vitamins, which aid in the establishment of extensive bacterial interactions. After being isolated using culture-dependent methods, bacterial symbionts were identified using morphological, biochemical, and molecular techniques. Functional screening was used to assess nutrient solubilization, nitrogen fixation, siderophore generation, ACC deaminase activity, and indole-3-acetic acid synthesis. Identifying these interactions promotes the development of long-term applications in algal biotechnology and environmental control, as well as providing useful insights into algal–bacterial symbiosis. This chapter emphasizes the isolation, identification, and functional properties of bacteria found in the *Chlorella* phycosphere.

Key words Photosphere, *Chlorella*, Bacterial, Algal–bacterial symbiosis

1 Introduction

Microalgae are the primary source of productivity in aquatic environments and are unicellular microorganisms that either directly or indirectly affect the global carbon cycle. They can produce and maintain a variety of high-value metabolites, making them incredibly beneficial in food, medicine, energy, and fisheries. Microalgae and bacteria coexist in both natural and artificial environments. During growth and reproduction, microalgae continuously release different metabolites into their surroundings, forming a microenvironment known as the phycosphere [5].

The relationship between eukaryotic and prokaryotic bacteria has been widely investigated for many decades, particularly those involving protists and bacteria. More recently, there has been an increased interest in the relationships and evolutionary interactions of bacteria and eukaryotic unicellular microalgae. Bell and Mitchell

Dhanasekaran Dharumadurai (ed.), *Chlorella Genomics, Cultivation, and Bioenterprise: Green Economy through Algal Innovations*, Springer Protocols Handbooks, https://doi.org/10.1007/978-1-0716-5276-3_7,
© The Author(s), under exclusive license to Springer Science+Business Media, LLC, part of Springer Nature 2026

refer to the "phycosphere," the area around algal cells where bacteria feed on the algae's extracellular products, as a niche in the environment inhabited by phytoplankton-associated bacteria. There are specialized interactions like communion or parasitism between bacteria and phytoplankton in the phycosphere, but there are also other kinds of interactions like rivalry, commensalism, or mutualism [4]. During photosynthesis, microalgae frequently excrete organic matter, including carbohydrates, nitrogen, organic acids, lipids, vitamins, biotin, proline, and glycolic acid, which have also been reported as extracellular organic matter released by *Chlorella*, thereby facilitating the survival of associated microbes.

In this symbiosis, connected bacteria provide vital nutrients such as nitrogen and phosphorus to enable algal growth [3]. The beneficial interactions between microalgae and bacteria have been widely utilized in wastewater treatment systems. Interactions between different species may increase nitrogen (N)- and phosphorus (P)-removal efficiency in the environment when compared to traditional single-step and multi-step processes. This might be achieved by merging the remarkable power of microorganisms to breakdown organic contaminants with algae's ability to absorb nitrogen, phosphorus, and other nutrients from wastewater. When combined with other treatment strategies, the symbiotic link between the carbon dioxide–oxygen cycle and the nutrient cycle established by algal–bacterial symbiosis may help in the development of a new technology to address the problem of excessive N, P, and organic matter in urban sewage water [2]. Bacteria can boost microalgal productivity and biomass yield through growth-promoting actions. However, the ability of certain bacteria to prevent algal growth has been employed as a biological strategy to control harmful algal blooms in aquatic environments [1].

2 Materials

1. Actively growing *Chlorella* culture.
2. Sterile distilled water/phosphate-buffered saline (PBS).
3. Petri plate.
4. Nutrient agar and nutrient broth.
5. Pikovskaya's (PVK) agar.
6. Potassium solubilizing bacteria agar.
7. Chrome azurol sulfonate (CAS) agar.
8. Azotobacter agar.
9. ADF agar supplemented with 1-aminocyclopropane-1-carboxylic acid (ACC).
10. Salkowski reagent.

11. Gram staining reagents.
12. Biochemical test reagents.
13. Ethanol (70%).

3 Methods

1. Harvest 10 mL of *Chlorella* culture during the mid-exponential phase (approx. day 5–7).
2. Centrifuge the *Chlorella* culture, discard the supernatant, and wash them twice with PBS to separate the symbiotic bacteria.
3. The sample was resuspended using sterile saline or PBS in a serial dilution series (10^{-1} to 10^{-6}).
4. Inoculate 100 μL of each dilution onto the Nutrient agar plates. Use a sterile L-spreader to evenly disperse the inoculum. The plates should be incubated inverted at 28 °C, which is the same temperature as the algal culture.
5. Observe the growth after 3 to 5 days. Choose colonies that differ in morphology (color, size, shape, border, and visibility). To obtain pure cultures, streak each distinct colony onto a new agar plate (Fig. 1).
6. Inoculate the culture with Pikovskaya's (PVK) agar, Potassium Bacteria Agar, and CAS agar to determine whether the culture is capable of producing phosphate and potassium solubilization as well as siderophores.

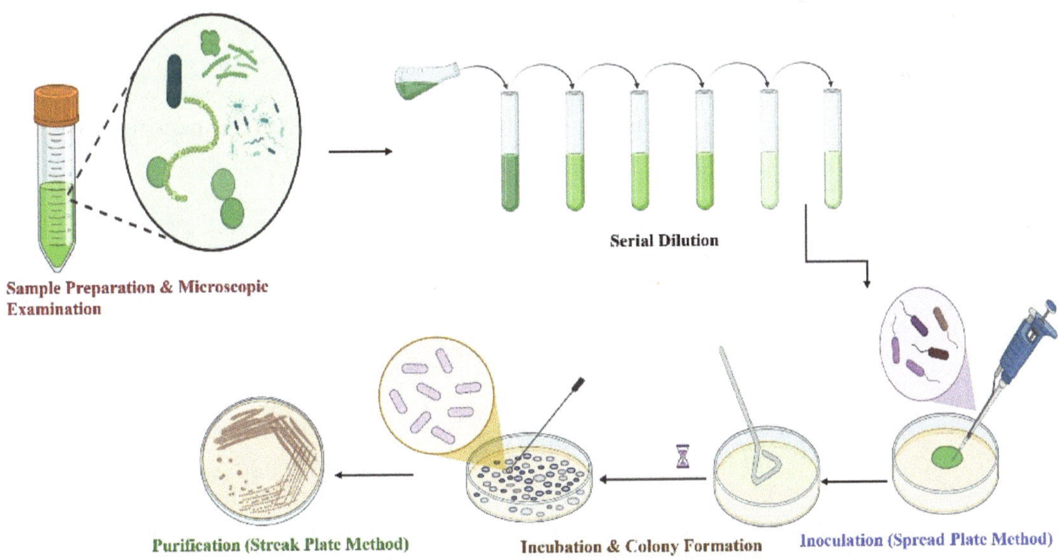

Fig. 1 Workflow for isolation and purification of bacterial symbionts from the *Chlorella* phycosphere

Table 1
Functional screening assays for bacterial symbionts from the *Chlorella* phycosphere

Target activity	Medium/reagent	Positive reaction indicator
Phosphate solubilization	Pikovskaya's (PVK) agar	Clear halo zone
Potassium solubilization	Potassium bacteria agar	Clear halo zone
Siderophore production	CAS agar	Orange halo zone
Nitrogen fixation	Azotobacter agar	Bacterial colony growth
ACC deaminase	ADF agar (with ACC)	Bacterial colony growth
IAA production	Salkowski reagent	Solution turns pink

7. Nitrogen fixation and ACC(1-Aminocyclopropane-1-carboxylate) deaminase activity were evaluated using Azotobacter agar and ADF agar.
8. Use Salkowski reagent to determine IAA production.

3.1 Observations

Cultures of *Chlorella* collected in the middle of the exponential phase were vibrantly green and healthy. After plating, functional symbionts of *Chlorella* phycosphere were identified by observations (Table 1).

4 Notes

1. Gentle washing approaches are necessary to maintain closely related bacterial symbionts.
2. All medium and reagents must be sterilized to avoid contamination.
3. The optimal incubation temperature and duration should promote bacterial growth without negatively impacting viability.

Data Availability Statement All data generated or analyzed during this study are provided in the chapter and its supplementary information files.

Ethics Approval Not applicable.

Author Contributions P.R., Writing original draft, Data analyses and methodology: D.D, Supervision, Writing—review and editing. All contributors read the Chapter and provided edits. All of the authors conceptualized the research. The chapter was read and approved by every author.

Competing Interests The authors declare that they have no competing financial or nonfinancial interests that could have influenced the work reported in this chapter.

Declaration of Generative AI and AI-Assisted Technologies in the Writing Process During the preparation of this work, the author(s) used AI to improve the readability of the chapter. After using this tool, the author(s) reviewed and edited the content as needed and take(s) full responsibility for the content of the publication.

Funding

Department of Biotechnology (Govt. of India) for the establishment of the National Repository for Microalgae and Cyanobacteria—Freshwater & Marine (NRMC F & M) (Phase II) (BT/PR29901/PBD26/694/2018) research. Chief Minister Research Grant (CMRG) for the financial support of research by the Government of Tamil Nadu, India, for the financial support (CMRG 2023–2024- Project No. CMRG2023BMBO7047).

References

1. Guo Z, Tong YW (2014) The interactions between Chlorella vulgaris and algal symbiotic bacteria under photoautotrophic and photoheterotrophic conditions. J Appl Phycol 26(3): 1483–1492
2. Ji X, Jiang M, Zhang J, Jiang X, Zheng Z (2018) The interactions of algae-bacteria symbiotic system and its effects on nutrients removal from synthetic wastewater. Bioresour Technol 247: 44–50
3. Rajendran PJ, Nooruddin T, Dharumadurai D (2025) Metabarcoding of bacterial communities impacting the dynamics of cyanobacterial blooms in freshwater reservoirs, South India. Water Air Soil Pollut 236(1):66
4. Serra V, Gammuto L, Petroni G, Ciurli A, Chiellini C (2022) Identification of the bacteria associated to the phycosphere of the chlorella-like strain SEC_LI_ChL_1. Algal Res 67:102869
5. Xiao L, Chen Z, Yang Y, Liu Z (2022) Growth promotion of Chlorella by symbiotic bacteria under adverse environments. Algal Res 66: 102799

Chapter 8

Isolation and Identification of Actinobacteria Associated in the Phycosphere of *Chlorella*

Subramaniyan Pasupathi and Dhanasekaran Dharumadurai

Abstract

The interactions between microalgae and bacteria are fundamental in aquatic environment. The term *phycosphere* is the microenvironment immediately surrounding an algal cell. Within the phycosphere, interactions between bacteria and microalgae exhibit specific interactions, which may be either synergistic or antagonistic. *Chlorella* is a unicellular freshwater green microalga that belongs to the class Chlorophyceae. It plays a significant role in carbon fixation and oxygen production in aquatic habitats such as ponds, lakes, rivers, dams, and oceans. Actinobacteria represent a Gram-positive bacterium that is widely distributed in terrestrial and aquatic ecosystem. Actinobacteria play an imperative part in shaping microbial community and influencing algal growth and enhancement of biomass production. Therefore, the present protocol aims to focus on isolation and identification of actinobacteria associated in the phycosphere of *Chlorella*.

Key words Phycosphere, Actinobacteria, *Chlorella*, Synergistic activity, Co-evolution, Aquatic habitat

1 Introduction

The interactions between eukaryotic and prokaryotic microorganisms have been comprehensively studied for several years, particularly those involving bacteria and microalgae. Attention has been focused on the association and co-evolution of bacteria and microalgae. The word 'phycosphere' refers to the immediate microenvironment surrounding the algal cell. In phycosphere, diverse interactions occur between bacteria and microalgae, including symbiotic or antagonistic activities. Microalgae are primary producers in aquatic ecosystem and play crucial role in global carbon cycling and nutrient dynamics. Phycosphere is enriched with organic compounds released through algal metabolism, including carbohydrate, amino acids, and vitamins. This zone supports dense and metabolically active bacterial populations and serves as a key site for microalgae–bacteria interactions [1]. Bacteria inhabiting the

Dhanasekaran Dharumadurai (ed.), *Chlorella Genomics, Cultivation, and Bioenterprise: Green Economy through Algal Innovations*, Springer Protocols Handbooks, https://doi.org/10.1007/978-1-0716-5276-3_8,
© The Author(s), under exclusive license to Springer Science+Business Media, LLC, part of Springer Nature 2026

phycosphere influence algal physiology by regulating nutrient availability, producing growth-promoting substance, and modulating stress responses. Microalgae provide organic substance that support bacterial growth, resulting in synergistic or antagonistic interactions [2].

Actinobacteria constitute one of the most abundant and ecological important bacterial species in freshwater and marine water ecosystem. They are known for their metabolic versatility, resilience to environmental stress, and ability to produce wide array of secondary metabolite, including antibiotics, enzymes, and signaling molecules [3].

The green microalga *Chlorella* is widely distributed in freshwater environment and is unicellular microalgae belonging to the class of Chlorophyceae. It is rich in protein, lipids, vitamins, and pigments, supporting its application in nutraceuticals, biofuel production, and wastewater treatment [4]. In aquatic ecosystem, *Chlorella* contributes to primary productivity and interacts with associated microorganisms within the phycosphere influencing nutrients [5]. Bold's Basal Medium (BBM) is a standard inorganic nutrient solution for culturing freshwater algae, consisting of macronutrient stock solutions ($NaNO_3$, $CaCl_2$, $MgSO_4$, K_2HPO_4, KH_2PO_4, NaCl) and trace element/EDTA/iron stocks, providing essential nitrogen, phosphorus, potassium, etc., crucial for algal growth.

Despite its importance, limited information is available on the diversity and functional potential of actinobacteria associated with the phycosphere of *Chlorella*. Understanding these associations may provide insights into positive or negative interactions of actinobacterial community in the phycosphere of *Chlorella*. Therefore, this protocol illustrates the requirements and general procedure involved in separation of phycosphere from the *Chlorella* pure culture and isolation and identification of actinobacteria associated in the phycosphere of *Chlorella*.

2 Materials

1. Sterile container.
2. Sterile conical flask.
3. Test tubes.
4. Bold's Basal Medium (BBM).
5. Vortex mixer.
6. Light microscope.
7. BOD incubator.
8. Phosphate-Buffered saline (PBS).

9. Starch Casein Agar (SCA).
10. Gram Staining Kit.
11. Catalase reagent.
12. Oxidase reagent.
13. Casein.
14. PCR master mix kit.
15. PCR equipment.
16. Agarose gel.
17. Electrophoresis unit.
18. Spectrophotometer.
19. *16S rRNA* universal primers.
20. Ulta deep freezer.
21. Test organisms.
22. Centrifuge tube.

Composition of Starch Casein Agar

Ingredients	Gms/Litre
Soluble starch	10.00
Casein (Vitamin free)	0.30
KNO_3	2.00
$MgSO_4.7H_2O$	0.05
K_2HPO_4	2.00
NaCl	2.00
CaCO3	0.02
$FeSO_4.7H_2O$	0.01
Agar	20.00

3 Methods

3.1 Sample Collection and Maintenance of Chlorella Cultures

1. Aseptically collect the freshwater sample from the ponds, lakes, river, and dam.
2. Transfer the algal sample into sterile container and transport to the laboratory [6].
3. Aseptically inoculate 10 ml of water sample into 200 ml of BBM (Bold basal medium) in the 500 ml conical flask.

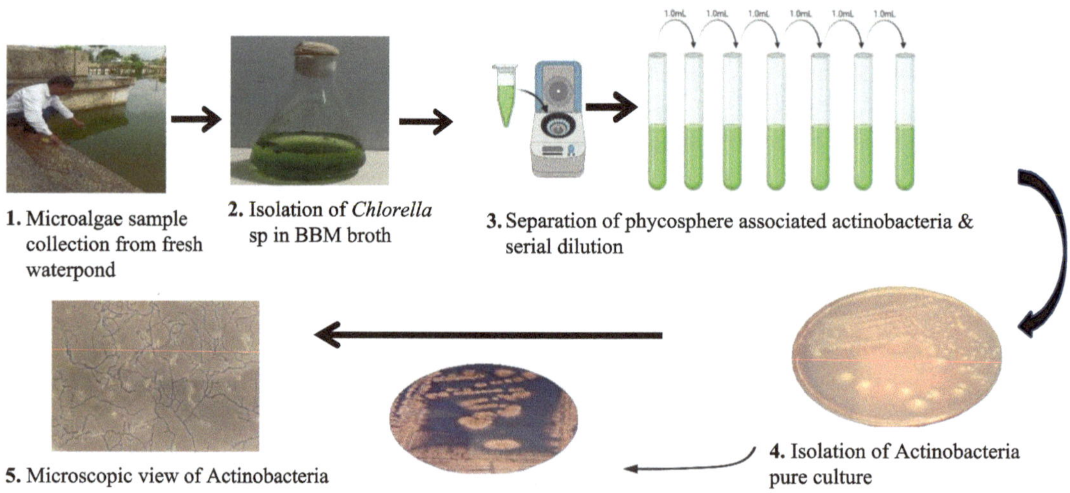

Fig. 1 Steps involved in the isolation of actinobacteria associated in the of *Chlorella* phycosphere

4. Incubate the culture at 28 °C on rotary shaker at 150 rpm under continuous illumination for 2–3 weeks.
5. Incorporate the antibacterial (nystatin) and antifungal (cycloheximide) antibiotics to culture medium to inhibit unwanted bacterial and fungal growth while allowing the growth of desired an axenic culture of *Chlorella* [7].
6. Monitor the Algal growth in every two days interval using an optical microscope.
7. Observe growth of algae in the 50 μl of algal culture spread-plate on BBM with agar and incubate at 28 °C for 2 weeks under continuous illumination.
8. Observe the colonies showing typical *Chlorella*-like morphology (small, circular, green, unicellular appearance), pure colony is streaked on BBM agar plates to obtain axenic culture (Fig. 1).
9. Obtain the pure *Chlorella* culture and identify the morphology under microscopic observation [6].

3.2 Separation of phycosphere from the Chlorella *Culture*

1. Inoculate *Chlorella* sp. into 200 ml BBM broth in the 500 ml conical flask and incubate for the period of 4 weeks to develop high concentration of symbionts in *Chlorella* culture.
2. Obtain 1 ml of aliquot *Chlorella* culture and centrifuge at 6000 rpm for 5 minutes at 4 °C.
3. After centrifugation, the suspension is washed twice with sterile distilled water to remove loosely attached bacteria.
4. Separate the phycosphere-associated bacteria from algal cell using the 0.22 μm membrane filter.
5. Inoculate the filter paper into 10 ml of sterile PBS and vortexed for 2–3 minutes to separate phycosphere-associated actionobacteria [7].

3.3 Isolation of Actinobacteria from the Phycosphere

1. Transfer 1 ml of supernatant into 9 ml of PBS and perform the serial up the 10^9 dilution.
2. Transfer 0.1 ml of serially diluted inoculum (10^{-6}, 10^{-7}, 10^{-8}) in Starch Casein Agar (SCA) by standard plant count method.
3. Using sterile L rod, spread the inoculum in the SCA culture plate by spread plate techniques.
4. Incubate the inoculated SCA plates culture at 28 °C ± 2 °C for 7–10 days.
5. After incubation, colonies that display typical actinobacterial characteristics (dry, chalky, filamentous, branching) are subcultured and this step is repeated 2 or 3 times.
6. Maintain the isolated pure cultures in SCA slants at 4 °C for further use.

3.4 Morphological Characterization and Biochemical Identification

1. Perform the morphological characterization using the Gram staining method (Gram- positive) and the stained bacterial cells are examined under light microscope.
2. Biochemical identification is performed using the catalase, oxidase, and casein hydrolysis tests.
3. The Gram-positive filamentous or branched mycelial structure, catalase-positive and oxidase-negative.
4. Casein hydrolysis, clear zone appears on skimmed milk agar.

3.5 Molecular Identification of Actinobacteria

3.5.1 Genomic DNA Extraction

1. Collect a small number of typical colonies and suspend in 200 μl of sterile water, the suspension centrifuge at 12000 rpm for 5 minutes at 4 °C.
2. Resuspend the pellet in 2 ml of lysis buffer (25 mM, pH 7.5) 10 mM Tris–HCL, 50 mM EDTA, and 0.1 M NaCl.
3. Mix the suspension thoroughly and this is subjected to five freeze-thaw process.
4. Add the Lysozyme (30 mg/mL) and RNAase (10 mg/mL) in microfuge tubes and incubate at 37 °C for 2 hours in shaking incubator.
5. After the incubation period, add the 20% SDS and Proteinase K (10 mg/mL).
6. Keep the tubes at 37 °C for 1 h to complete cell lysis.
7. Centrifuge the lysate at 13,000 rpm at 25 °C for 12 minutes.
8. Carefully transfer the supernatant containing genomic DNA to a fresh centrifugation tube.
9. Store the extracted gDNA samples at −20 °C.

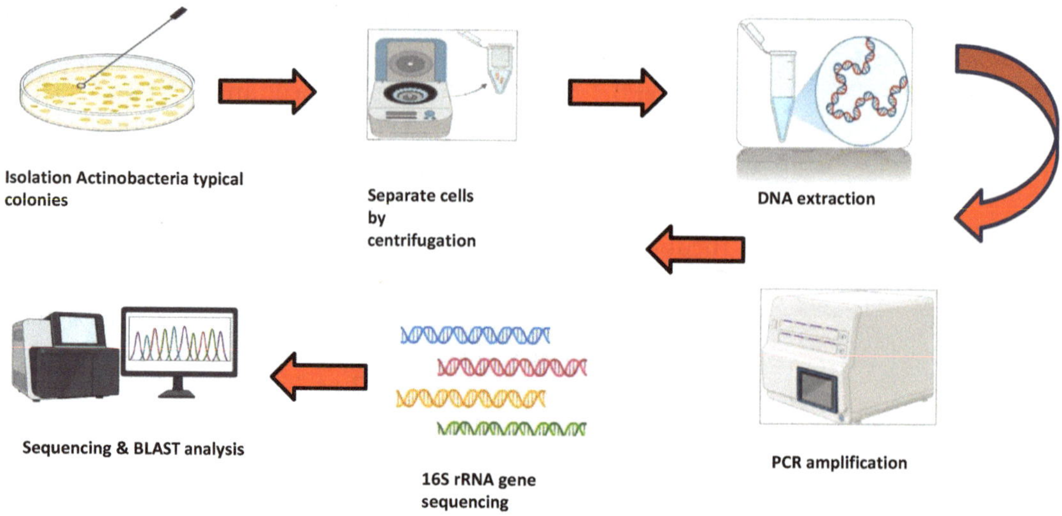

Fig. 2 *16S rRNA* gene sequencing of actinobacteria

3.5.2 Quality Control

1. Quantity gDNA determine by UV absorbance of the DNA measured by using a Nanodrop spectrophotometer.
2. Asses the quality of the DNA using 1% agarose gel electrophoresis at 100 V for 45 min, and record the OD value at 260/280 ratio at 560 nm by using spectrophotometer [8].

3.5.3 Polymerase Chain Reaction (PCR)

1. Perform the PCR amplification by the 16S rRNA primers design to amplify the region between 27 and 765.
2. Use the primer 27f (5-AGAGTTTGATCMTGGCTCAG) and 765r (5- CTGTTTGCTCCCCACGCTTTC).
3. Purify the amplified PCR product using a PCR Prep DNA purification kit (Promega) and sequencing is performed using the 765r primer.
4. *16S rRNA* gene is performed using 27f, 765r, 704f (5′-GTA GCGGTGAAATGCGTAGA- 3′) and 1492r primers to obtain near full-length *16S rRNA* gene sequence (Fig. 2).
5. Compare the obtained *16S rRNA* gene sequences with sequences available in the GenBank database using the BLASTN program for molecular identification (*see* **Notes 1–4**) [9].

4 Notes

1. Diversity of actinobacteria inhabiting the phycosphere of *Chlorella* is documented.
2. Limited information is available on the interaction between *Chlorella* and associated actinobacteria.

3. Integrated morphological, biochemical, and molecular characterization of actinobacteria from the *Chlorella* phycosphere remains insufficiently addressed.
4. Symbiotic or antagonistic interactions of actinobacteria in the phycosphere will be recorded in *Chlorella* biomass production and algal bloom control, respectively.

References

1. Seymour JR, Amin SA, Raina JB, Stocker R (2017) Zooming in on the phycosphere: the ecological interface for phytoplankton–bacteria relationships. Nat Microbiol 2(7):1–12
2. Amin SA, Parker MS, Armbrust EV (2012) Interactions between diatoms and bacteria. Microbiol Mol Biol Rev 76(3):667–684
3. Barka EA, Vatsa P, Sanchez L et al (2016) Taxonomy, physiology, and natural products of Actinobacteria. Microbiol Mol Biol Rev 80(1):1–43
4. Spolaore P, Joannis-Cassan C, Duran E, Isambert A (2006) Commercial applications of microalgae. J Biosci Bioeng 101(2):87–96
5. Ramanan R, Kim BH, Cho DH, Oh HM, Kim HS (2016) Algae–bacteria interactions: evolution, ecology and emerging applications. Biotechnol Adv 34(1):14–29
6. Rajendran PJ, Shunmugam S, Saravanan C, Nooruddin T, Dharumadurai D (2025) Molecular monitoring, identification of the microcystin synthetase (mcy) gene and toxic potential of *Microcystis aeruginosa* in Indian freshwater blooms. Toxicon:108631
7. Priya Dharsini P, Dhanasekaran D, Gopinath PM, Ramanathan K, Shanthi V, Chandraleka S, Biswas B (2017) Spectroscopic identification and molecular modeling of diethyl 7-hydroxytrideca-2, 5, 8, 11-tetraenedioate: a herbicidal compound from *Streptomyces* sp. Arab J Sci Eng 42(6):2217–2227
8. Rajendran PJ, Nooruddin T, Dharumadurai D (2025) Metabarcoding of bacterial communities impacting the dynamics of cyanobacterial blooms in freshwater reservoirs, South India. Water Air Soil Pollut 236(1):66
9. Coombs JT, Franco CM (2003) Isolation and identification of actinobacteria from surface-sterilized wheat roots. Appl Environ Microbiol 69(9):5603–5608

Chapter 9

Separation of Amplified Gene Using Agarose Gel Electrophoresis

Prakash Palanivel, Priyanka Jayam Rajendran, Dhanasekaran Dharumadurai, and Gangatharan Muralitharan

Abstract

Agarose gel electrophoresis is a basic molecular biology technique that separates and visualizes DNA fragments via size-dependent migration through an agarose gel. A methodical approach to gel preparation, agarose concentration selection, sample loading, electrophoresis settings, and DNA visualization using blue-light or UV systems is presented in this chapter. Important factors that impact fragment resolution, such as buffer selection, voltage, and stain selection, are discussed in order to attain optimal efficiency. Common uses for this technique include verifying PCR amplification, assessing restriction digestion, testing DNA quality, and separating particular DNA fragments for additional molecular analysis. Because of its affordability, ease of use, and repeatability, it is a vital instrument for research and educational facilities, providing reliable and excellent results in standard molecular biology techniques.

Key words Agarose gel electrophoresis, PCR Products, DNA visualization, DNA fragment separation, Blue-light/UV transillumination, Electrophoresis

1 Introduction

Agarose gel electrophoresis is a common technique used in molecular biology to separate and visualize nucleic acids such as DNA and RNA. This approach is based on the concept that when an electric field is provided, DNA molecules migrate to the positive electrode due to the negative charge on their phosphate backbone. During electrophoresis, the agarose gel functions as a filter, allowing smaller fragments to flow through faster and bigger fragments to move slower [1]. This size distinction allows for more accurate determination of DNA fragment lengths than a molecular weight marker.

DNA fragment resolution is determined by several parameters, including agarose level, buffer composition, voltage, and run time. In general, lower amounts of agarose are more successful at separating bigger DNA fragments. Common buffers, which include

Dhanasekaran Dharumadurai (ed.), *Chlorella Genomics, Cultivation, and Bioenterprise: Green Economy through Algal Innovations*, Springer Protocols Handbooks, https://doi.org/10.1007/978-1-0716-5276-3_9,
© The Author(s), under exclusive license to Springer Science+Business Media, LLC, part of Springer Nature 2026

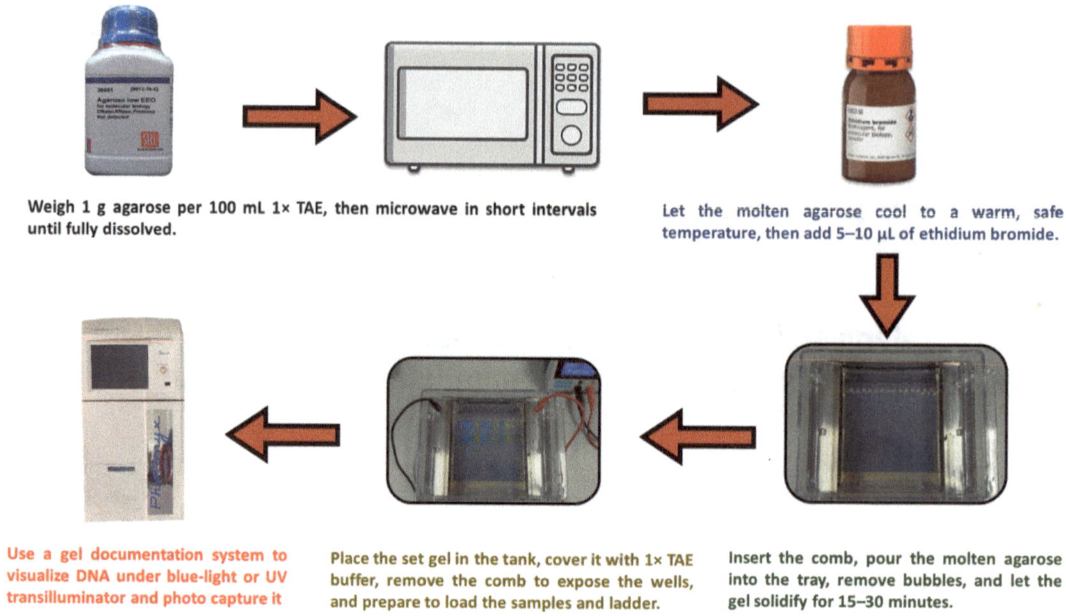

Fig. 1 Schematic representation of agarose gel preparation, casting, electrophoresis, and visualization of DNA using a gel documentation system

Tris-Acetate-EDTA (TAE) or Tris-Borate-EDTA (TBE), help to maintain pH stability while also providing the ions needed for current flow. DNA bands can be observed with intercalating dyes such as SYBR Safe, GelRed, and Ethidium Bromide when exposed to ultraviolet (UV) or blue light [2].

Agarose gel electrophoresis is a key step in determining the ability of polymerase chain reaction (PCR) amplification [3]. It allows researchers determine whether a specific gene segment was amplified and whether primer dimers or nonspecific amplifications exist [4]. Furthermore, the procedure evaluates the efficiency of restriction enzyme digestion, examines DNA quality, and purifies specific DNA fragments for use in cloning, sequencing, or mutagenesis [5]. Because of its simplicity, low cost, and dependability, agarose gel electrophoresis is an essential tool in both academic and research labs. This chapter presents a detailed approach for separating and visualizing PCR-amplified fragments using the agarose gel electrophoresis technique (Fig. 1).

2 Materials

1. Agarose powder.
2. 10× TAE buffer.
3. DNA ladder (e.g., 100 bp ladder, 1 kb ladder).

4. PCR product.
5. 6× DNA loading dye.
6. Ethidium Bromide.
7. Gel casting tray, comb, electrophoresis chamber, power supply.
8. Microwave or hotplate.
9. Heat resistant flask or bottle.
10. Measuring cylinder and spatula.
11. Blue-light transilluminator or UV transilluminator and gel imaging device.

3 Methods

3.1 Choose Gel Percentage

1. Based on the expected fragment size, choose the agarose concentration from the Table 1.

3.2 Prepare and Cast the Gel (for Instance, 100 mL of 1% Gel)

1. Weigh 1.0 g agarose to 100 mL of 1× TAE in a heat-resistant flask.
2. Heat the mixture in a microwave in short intervals, until the agarose fully dissolves and the solution becomes clear.
3. Allow the molten agarose to cool until it is warm but safe to handle.
4. Carefully add 5–10 μL of Ethidium Bromide to the molten agarose.
5. Place the comb into the gel casting tray and pour the molten agarose into it.
6. Use a pipette tip to gently remove any bubbles.
7. Let the gel solidify at room temperature for 15–30 minutes.

3.3 Set up the Electrophoresis Chamber

- Place the solidified gel in the electrophoresis tank.
- Add enough 1× TAE running buffer to cover the gel by 2–3 mm.

Table 1
Agarose gel concentration and suitable DNA fragment size range

Agarose gel (%)	Suitable fragment size
0.5–0.8%	Very large fragments (>1.5 kb)
1.0%	General purpose (500–5000 bp)
1.5%	Medium fragments (200–1000 bp)
2.0% or higher	Small fragments (100–500 bp)

- Make sure the wells are immersed but the comb cavities are not flooded.
- Carefully remove the comb to reveal the wells.

3.4 Prepare Samples and Ladder

- Add 5 volumes of the PCR sample with 1 volume of 6× loading dye to get a final concentration of 1×.
- Prepare the DNA ladder according to the manufacturer's instructions, then dilute it as needed.
- To visualize a large PCR volume, pipette an aliquot of 1–10 μL or a defined amount in ng if quantified.

3.5 Load Gel and Run Electrophoresis

- Load the ladder into the first lane, followed by samples in the remaining wells while keeping track of sample IDs.
- Close the lid and connect the electrodes and let DNA migrate to the positive electrode.
- Set voltage according to gel size and desired resolution (e.g., between 50 and 100 V).
- Start the run and monitor the dye front migration.
- Stop electrophoresis when the tracking dye reaches about 75 to 85% of the gel length.
- Typical run times are 20 to 60 minutes, depending on the voltage and gel percentage.

3.6 Visualize and (Optionally) Excise Bands

- Turn off the power and gently remove the gel from the tank.
- Use a gel documentation system to visualize DNA under blue-light or UV transilluminator and photo capture it .
- To purify a band, minimize UV exposure, excise the band with a clean scalpel, and proceed with a gel extraction kit following the kit protocol.

3.7 Expected Observations

The PCR-amplified gene fragments were successfully separated by the agarose gel electrophoresis experiment. A 1.5% agarose gel was prepared with 1× TAE buffer and ethidium bromide. Once the gel was solidified, the PCR products and DNA ladder was loaded into the wells. Electrophoresis was carried out at 100 V for approximately 45 minutes. The bromophenol blue tracking dye indicated a constant current flow across the gel during the run by moving continuously in the direction of the anode.

After electrophoresis, the gel was observed with a blue light transilluminator. Most of the lanes revealed distinct DNA bands (Fig. 2). The well-resolved bands on the molecular weight marker suggested optimal gel activity. Each PCR sample exhibited a single, distinct band that matched the expected amplicon size when compared to the DNA ladder, demonstrating target gene-specific amplification. Light secondary bands and smearing were observed

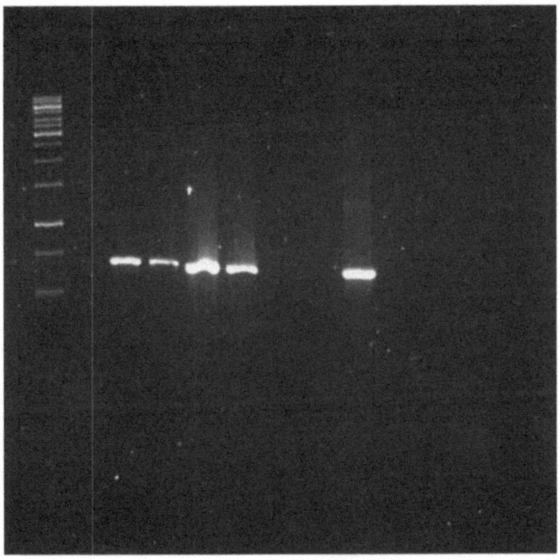

Fig. 2 Agarose gel electrophoresis image showing DNA bands separated on a 1% agarose gel and visualized under UV/blue-light illumination

in a few lanes, which could be attributed to primer-dimer formation or nonspecific amplification. Overall, the results showed that agarose gel electrophoresis was an effective approach for distinguishing amplified gene fragments.

4 Notes

1. Always handled EtBr in a chemical fume hood or a specialized area.
2. Dispose of EtBr waste (gels, tips, tubes) in biohazard or chemical waste bins marked for mutagenic materials.
3. Ensure proper tank assembly and correct buffer level to avoid short circuits.
4. When using a UV transilluminator, wear a face shield or UV-protective goggles [6].

References

1. Arakawa T, Nakagawa M, Sakuma C, Tomioka Y, Kurosawa Y, Akuta T (2024) Polysaccharide as a separation medium for Gel Electrophoresis. Polysaccharides 5(3):380–398
2. Pandey A, Momeni O, Pandey P (2024) Quantitative analysis of genomic DNA degradation of E. coli using automated gel electrophoresis under various levels of microwave exposure. Gels 10(4):242

3. Palanivel P, Rajendran PJ, Dharumadurai D, Gangatharan M (2024) Colony PCR for rapid detection and identification of microalgae. In: Plant microbiome engineering. Springer US, New York, pp 243–247
4. Yang J, Zhang T, Yang B, Liu J, Li Z, Yamaguchi Y (2025) High-resolution analysis of DNA electrophoretic separations via digital image processing. Separations 12(11):296
5. Martin R (2020) Gel electrophoresis: nucleic acids. Garland Science
6. Sambrook J, Fritsch EF, Maniatis T (1992) Molecular cloning: a laboratory manual II. New York, Chicago

Chapter 10

PCR Amplification of 18S rRNA Gene in *Chlorella* sp.

Yingjie Liu, Ning Xu, Hongwei Chai, and Peng Chen

Abstract

The 18S rRNA gene serves as a conserved ribosomal RNA sequence in eukaryotic algae, including *Chlorella* species, making it an ideal internal reference for quantitative PCR (qPCR) assays aimed at normalizing the abundance of target genes involved in genomics, cultivation optimization, and bioenterprise applications. In *Chlorella* genomic DNA, the 18S rRNA gene enables relative quantification of variable target fragments, such as those encoding lipid biosynthesis enzymes or stress-response proteins, by providing a stable baseline for expression analysis. This protocol details the SYBR Green-based qPCR amplification of the 18S rRNA gene from *Chlorella* genomic DNA, facilitating accurate, real-time detection and quantification. The method is rapid, cost-effective, and suitable for high-throughput screening in algal research, supporting green economy initiatives through enhanced algal strain monitoring. Key steps include DNA extraction, reaction setup, thermal cycling, and data analysis using the $2^{-\Delta\Delta Ct}$ method. This approach ensures reliable amplification efficiency (90–110%) and specificity, which can be confirmed by melting curve analysis.

Key words qPCR, SYBR Green, 18S rRNA, *Chlorella*, Algal genomics, Relative quantification

1 Introduction

Chlorella, a genus of green microalgae, holds significant promise in bioenterprise due to its rapid growth, high biomass productivity, and applications in biofuel production, nutraceuticals, and wastewater remediation [1]. Genomic studies of *Chlorella* reveal diverse metabolic pathways that underpin these traits, but accurate quantification of gene expression requires robust normalization controls. The 18S rRNA gene, a highly conserved eukaryotic ribosomal RNA sequence spanning approximately 1800 base pairs, exhibits stable copy numbers across *Chlorella* strains, rendering it an optimal housekeeping gene for qPCR-based relative quantification [2]. Unlike variable target genes (e.g., those involved in lipid accumulation or CO_2 fixation), the 18S rRNA gene's conservation allows it to serve as an internal reference, enabling the calculation of fold changes in target fragment abundance via the $2^{-\Delta\Delta Ct}$ method [3]. Traditional detection methods for algal gene expression, such

Dhanasekaran Dharumadurai (ed.), *Chlorella Genomics, Cultivation, and Bioenterprise: Green Economy through Algal Innovations*, Springer Protocols Handbooks, https://doi.org/10.1007/978-1-0716-5276-3_10,
© The Author(s), under exclusive license to Springer Science+Business Media, LLC, part of Springer Nature 2026

as Northern blotting or endpoint PCR, lack the sensitivity and real-time capabilities of qPCR [4]. Advances in molecular biology position SYBR Green-qPCR as a preferred tool for *Chlorella* research, offering exponential amplification monitoring through dsDNA-intercalating dyes [5]. In this protocol, the 18S rRNA gene is amplified alongside user-defined target primers to quantify relative expression levels, supporting applications in strain selection for cultivation and bioenterprise scalability. For instance, co-amplification with primers targeting the *acetyl-CoA carboxylase* gene (*accD*) allows assessment of lipid pathway activity relative to total ribosomal content [6]. The procedure employs standard thermal cycling conditions to achieve Ct values between 15 and 35, ensuring detection limits suitable for environmental or cultured samples. Post-amplification melting curve analysis confirms product specificity by verifying a single peak, indicative of primer efficiency [7]. This method aligns with post-pandemic advancements in qPCR infrastructure, promoting accessible algal innovations for a green economy [8].

2 Materials

1. RNase-free ddH_2O.
2. Bacterial (Algal) Genomic DNA Extraction Kit (e.g., Qiagen DNeasy Plant Mini Kit).
3. 6× DNA loading buffer.
4. DNA marker (e.g., 100–5000 bp ladder).
5. Agarose powder.
6. Microwave oven.
7. Nucleic acid dye (e.g., GelRed).
8. 1× TAE buffer.
9. Gel electrophoresis apparatus.
10. Gel imaging system.
11. NanoDrop spectrophotometer.
12. qPCR Master Mix kit (containing 2× SYBR Green Mix, ROX reference dye).
13. Forward and reverse primers for 18S rRNA gene (10 μM stock; sequences: F: 5′-GTAGTTCTAGGGACGAATGA-3′, R: 5′-TTTACTTTGTCACTACCTCT-3′ [9]).
14. Optional: Primers for target gene (e.g., 10 μM *accD* F/R for lipid quantification).
15. 96-well PCR tube rack.
16. Ice bucket or ice maker.

17. 0.2 mL individual PCR tubes or 8-strip PCR tubes with optical caps.
18. Pipettes (1–10 μL, 10–100 μL, 20–200 μL) with tips.
19. Mini centrifuge.
20. Real-time qPCR instrument (e.g., Applied Biosystems QuantStudio).

3 Methods

3.1 DNA Extraction from Chlorella, Quantitation, and Quality Control

1. Harvest 50–100 mg wet *Chlorella* biomass by centrifugation at 5000 × g for 5 min. Resuspend in lysis buffer from the extraction kit and follow the manufacturer's protocol for algal DNA isolation [10].
2. Elute DNA in 50 μL RNase-free ddH_2O and store at −20 °C until use. Prepare a 1.5% agarose gel: Dissolve 0.6 g agarose in 40 mL 1× TAE buffer by microwaving for 1–2 min until clear.
3. Cool slightly, add 4 μL GelRed, mix, and pour into the gel tray. Allow to solidify for 30 min. Load samples: In a PCR tube, mix 5 μL extracted DNA with 1 μL 6× loading buffer.
4. Load 3 μL DNA marker into the first well, followed by 6 μL sample mixtures into subsequent wells. Electrophorese at 100 V for 30 min in 1× TAE buffer.
5. Image the gel using a gel documentation system.
6. Assess quality by the presence of a single, bright high-molecular-weight band without smearing, indicating intact DNA [11].
7. Quantitate DNA: Dilute 1 μL sample in 1 μL ddH_2O and measure absorbance at 260 nm using NanoDrop.
8. Aim for concentrations of 20–100 ng/μL with A260/A280 ratios of 1.8–2.0 for purity.

3.2 Preparation of the qPCR System and Thermal Cycling

1. Assemble a 10 μL reaction volume per well (Table 1).
2. Use 0.3 μM final concentration for each primer pair (18S rRNA and optional target). The 2× SYBR Green Mix and 50× ROX dye are used at 1× final concentration. Prepare reactions on ice in a laminar flow hood to minimize contamination.
3. Set up triplicates for each sample, target (if applicable), and no-template controls (NTC).
4. For multi-gene quantification, prepare master mixes separately for 18S rRNA and target reactions.
5. To minimize pipetting errors, scale up volumes by 10–20% (e.g., for 3 wells: 15 μL DNA, 30 μL Mix + primers/ROX/

Table 1
qPCR system for 18S rRNA amplification

Reagent	Volume (μL)
2× SYBR green mix	5.0
10 μM forward primer (F)	0.3
10 μM reverse primer (R)	0.3
Template DNA (10 ng/μL)	1.0
50× ROX reference dye	0.2
RNase-free ddH_2O	3.2
Total	10.0

Table 2
Two-step reaction procedure of qPCR

Stage	Cycle	Temp (°C)	Time	Content
Initial denaturation	1×	95	15 min	Initial denaturation
PCR	40×	95 60–66	10 sec 20–32 sec	Denaturation Annealing/extension
Melting/dissociation curve stage				

Note: Melting curve analysis is only applicable to real-time amplified fragments that retain fluorescent groups. It can be used to detect the presence of nonspecific products

ddH_2O aliquot). Centrifuge 8-strip tubes briefly (3 s at 1000 × g) to collect droplets.

6. Load tubes into the qPCR instrument and run the following program (Table 2): Initial denaturation at 95 °C for 15 min; 40 cycles of 95 °C for 10 s (denaturation), and 60 °C for 30 s (annealing/extension); followed by melt curve analysis (95 °C for 60 s, 60 °C for 60 s, 95 °C for 60 s with 0.5 °C increments) [12].
7. Acquire fluorescence during the extension step using the SYBR channel.

3.3 Calculation of Experimental Results

1. Export Ct values from the qPCR instrument software. Calculate the mean Ct value for 18S rRNA triplicates (mCt_{18S}) and, if applicable, for target gene triplicates (mCt_{target}).
2. Determine ΔCt as mCt_{target} - mCt_{18S}, which quantifies the relative abundance of the target gene normalized to the 18S rRNA reference [3].

3. For inter-sample fold-change analysis, compute $\Delta\Delta Ct = \Delta Ct_{sample} - \Delta Ct_{calibrator}$, followed by fold change $= 2^{-\Delta\Delta Ct}$.
4. To assess amplification efficiency, prepare a standard curve from ten-fold serial dilutions (eight gradients) of a plasmid harboring the 18S rRNA amplicon (10 to 10^8 copies/μL).
5. Plot Ct against $\log_{10}$(copies) to obtain the slope (k), then derive efficiency $E = [10^{(-1/k)} - 1] \times 100\%$. Consider the reaction valid if E falls within 90–110% [13].

4 Notes

Expected outcomes include Ct values typically ranging from 18 to 25 for 10 ng *Chlorella* DNA input, accompanied by single melt curve peaks at approximately 82 °C to verify amplicon specificity. Relative quantification facilitates the assessment of fold changes in target gene expression—for instance, elevated levels in nutrient-stressed cultures relative to controls—when normalized to 18S rRNA.

- Perform all steps on ice and protect SYBR Green from light to maintain dye integrity.
- Employ a 10 μL reaction volume for cost-effective screening; increase to 20 μL for enhanced precision with low-abundance targets.
- Validate primers annually using BLAST against *Chlorella* genomes to confirm sequence conservation [14].
- The detection limit is approximately 10^2 copies per reaction; Ct values below 35 are deemed reliable.
- Select optical-grade plastics compatible with the qPCR instrument; autoclave pipette tips to ensure sterility.
- For high-biomass *Chlorella* cultures (> 10^7 cells/mL), dilute extracted DNA 1:10 to prevent PCR inhibition.

Acknowledgments

This work was supported by Gansu Provincial Science and Technology Major Project (Grant No. 24ZDFA001), The Lanzhou Municipal Science and Technology Program (Grant Nos. 2024-4-2, 2025-2-100 and 2025-2-103), Scientific Research Project of Gansu Medical Products Administration (Grant No. 2025GSMPA087), Open Project of NMPA Key Laboratory for Quality Control of Traditional Chinese Medicine (Grant No. 2025GSMPA-KL03), Gansu Intellectual Property Project (Grant No. 22ZSCQ065) and the College

Students' Innovation and Entrepreneurship Program of Lanzhou University, China (Grant Nos. 20250260006, 20250260016 and 20250260020).

References

1. Scott SA, Davey MP, Dennis JS, Horst I, Howe CJ, Lea-Smith DJ, Smith AG (2010) Biodiesel from algae: challenges and prospects. Curr Opin Biotechnol 21:277–286
2. Sogin ML, Elwood HJ, Gunderson JH (1986) Evolutionary diversity of eukaryotic small-subunit rRNA genes. Proc Natl Acad Sci 83: 1383–1387
3. Livak KJ, Schmittgen TD (2001) Analysis of relative gene expression data using real-time quantitative PCR and the $2^{-\Delta\Delta CT}$ method. Methods 25:402–408
4. Sambrook DRJ (2016) Molecular cloning: a laboratory manual, 3rd ed.
5. Arya M, Shergill IS, Williamson M, Gommersall L, Arya N, Patel HR (2005) Basic principles of real-time quantitative PCR. Expert Rev Mol Diagn 5:209–219
6. Rismani-Yazdi H, Haznedaroglu BZ, Bibby K, Peccia J (2011) Transcriptome sequencing and annotation of the microalgae *Dunaliella tertiolecta*: pathway description and gene discovery for production of next-generation biofuels. BMC Genomics 12:148
7. Ririe KM, Rasmussen RP, Wittwer CT (1997) Product differentiation by analysis of DNA melting curves during the polymerase chain reaction. Anal Biochem 245:154–160
8. Mullis KB (1990) The unusual origin of the polymerase chain reaction. Sci Am 262:56–65
9. Medlin L, Elwood HJ, Stickel S, Sogin ML (1988) The characterization of enzymatically amplified eukaryotic 16S-like rRNA-coding regions. Gene 71:491–499
10. Doyle JJ (1990) Isolation of plant DNA from fresh tissue. Focus 12:13–15
11. Brons JK, van Elsas JD (2008) Analysis of bacterial communities in soil by use of denaturing gradient gel electrophoresis and clone libraries, as influenced by different reverse primers. Appl Environ Microbiol 74:2717–2727
12. Bustin SA, Benes V, Garson JA, Hellemans J, Huggett J, Kubista M, Mueller R, Nolan T, Pfaffl MW, Shipley GL (2009) The MIQE guidelines: minimum information for publication of quantitative real-time PCR experiments, in, Oxford University Press,. 55:611
13. Pfaffl MW (2001) A new mathematical model for relative quantification in real-time RT–PCR. Nucleic Acids Res 29:e45–e45
14. Altschul SF, Gish W, Miller W, Myers EW, Lipman DJ (1990) Basic local alignment search tool. J Mol Biol 215:403–410

Chapter 11

18S rRNA Gene Sequencing and BLAST Analysis Using NCBI

Asraf Sithikka Rasheed, Omm Bhusan Sarangi, Thajuddin Nooruddin, and Dhanasekaran Dharumadurai

Abstract

Molecular assessment of microalgae is crucial for the biodiversity studies, reliable taxonomic evaluation, and further ecological and biotechnological applications. This protocol describes a comprehensive workflow for isolating *Chlorella* to obtain pure culture, extracting genomic DNA, amplifying the *18S rRNA* gene through PCR, obtaining high-quality bidirectional Sanger sequencing data, and performing comparative sequence analysis using the NCBI BLASTn tool. The technique enables effective identification of eukaryotic microalgae by concentrating on conserved and hypervariable sections of the gene. The workflow has been designed to be accessible, repeatable, and suitable for regular taxonomic analysis of multiple strains. The protocol outlines the crucial steps from sample preparation to sequence curation and also aid in the interpretations of the BLAST findings in order to achieve reliable species and strain level identification.

Key words *Chlorella*, Taxonomic analysis, *18S rRNA*, Sanger sequencing, NCBI BLASTn tool

1 Introduction

Molecular identification of microalgae is crucial to understand their taxonomy and also for a broad array of ecological and biotechnological applications. Sequencing provides rapid and efficient identification of algal species by addressing the significant challenges faced by the microscopic analysis, it involves fewer resources and professionals, can process hundreds of samples at once by using high-throughput, parallel techniques, and enables the identification of organisms lacking distinctive morphological traits [1]. *18S rRNA* is the primary structural and functional RNA constituent that forms the fundamental unit of the small ribosomal subunit in the cytoplasmic ribosomes of the eukaryotes. The gene that codes for *18S rRNA* is *18S rDNA* has been well recognized in phylogenetic studies which is also an essential marker for the randomized targeting PCR in the screening of environmental biodiversity [2]. The nuclear smaller subunit ribosomal RNA gene–*18S rRNA*

Dhanasekaran Dharumadurai (ed.), *Chlorella Genomics, Cultivation, and Bioenterprise: Green Economy through Algal Innovations*, Springer Protocols Handbooks, https://doi.org/10.1007/978-1-0716-5276-3_11,

© The Author(s), under exclusive license to Springer Science+Business Media, LLC, part of Springer Nature 2026

is an ideal gene marker for the eukaryotic microorganisms because of its stability, their conserved sections that enables facilitated targeting with the primers and hypervariable regions (V1 through V9) which offers sequence diversity for taxonomic classification [1, 3]. Recent studies evaluating variable regions throughout the complete *18S rRNA* gene that possess conserved regions that are ideal for priming the amplification of hypervariable regions in eukaryotes and eukaryotic plankton [1]. In addition to serving as the foundation for more recent automated methods, the conventional Sanger sequencing is still the most often used sequencing method for various purposes [4]. The Basic Alignment Search Tool (BLAST) is an online platform that aids in comparative analysis of genes through sequence similarity, where it finds sequences in a database that are similar to the user's query sequence. Distinct BLAST variants are available for the comparison of nucleotide or protein queries against corresponding nucleotide or protein databases [5]. This chapter provides a comprehensive step by step protocol for the *18S rRNA* gene sequencing and BLAST analysis.

2 Materials

1. DNA extraction kit.
2. PCR thermal cycler.
3. PCR mix.
4. 100 bp and 1 kb DNA ladder.
5. Gel electrophoresis system.
6. Gel documentation system.
7. NanoDrop or Qubit fluorometer.
8. ExoSAP-IT or PCR cleanup kit or gel extraction kit.
9. Sequencing primers.
10. BigDye Terminator v3.1 Cycle Sequencing Kit.
11. Chromatogram viewers.
12. Sequence assemblers.
13. Multiple sequence alignment tools.
14. NCBI BLASTn tool.

3 Methods

3.1 Axenic Culture Development and DNA Extraction

1. From the environmental samples, monoclonal axenic cultures of *Chlorella* must be developed through single-cell isolation or streak plating method on appropriate agar media. Wash the cells twice or thrice in distilled water and remove the media salts prior to the DNA extraction (*See* **Note 1**) [6].

2. Lyse the cell pellet in CTAB buffer (2% CTAB, 100 mM Tris-HCl pH 8.0, 1.4 M NaCl, 20 mM EDTA) with β-mercaptoethanol and incubate for about 30 to 60 minutes at 60 °C.
3. Extract the DNA with the solution of chloroform and isoamyl alcohol and precipitate with isopropanol. Wash the precipitate in 70% ethanol, air dry, and resuspend it in TE buffer or nuclease free water.
4. Qualify the integrity of the extracted DNA on 0.8 to 1% of agarose and quantify the concentration of DNA (NanoDrop or Qubit). A sufficient degree of purity is indicated by a spectrophotometric A260/280 ratio of around 1.8 (*See* **Note 2**) [6–9].

3.2 PCR Amplification

1. Combine the PCR mix with the total volume of 25 μL consisting of 1.0X PCR buffer, 1.0 μM of each forward and reverse primer, 0.2 mM dNTPs, 3 mM MgCl2, 2.5 U of Pfu DNA polymerase along with 3 μL of template DNA.
2. Perform the amplification in a thermal cycler with the following thermal profile, 95 °C—5 mins for denaturation; 35 cycles of 95 °C—30 secs, 50 °C—1 min, 72 °C—7 mins for annealing; and 72 °C for 7 minutes for extension.
3. Run 1.0% of agarose gel electrophoresis with 3 to 5 μL of PCR product along with 100 bp ladder and expect a clear band at expected size.
4. Purify the PCR product by ExoSAP–IT or column purification and quantify the concentration of the product (*See* **Note 3**) [3, 9, 10].

3.3 Gene Sequencing

1. For high confidence species identification, perform a standard workflow involving bidirectional Sanger sequencing to generate overlapping forward and reverse reads.
2. Visually inspect the chromatograms using software such as Chromas or FinchTV in order to remove the low-quality ends which are characterized by ambiguous and overlapping peaks.
3. Assemble the forward and reverse reads using sequence assembly software or alignment tools (*See* **Notes 4–5**) [4, 9].

3.4 BLAST Analysis

1. Go to NCBI BLAST webserver (https://blast.ncbi.nlm.nih.gov/Blast.cgi) for BLAST analysis and select the Nucleotide BLAST (blastn).
2. Paste/upload the aligned FASTA sequence in the query sequence of blastn, set the maximum target sequences and click BLAST to run.

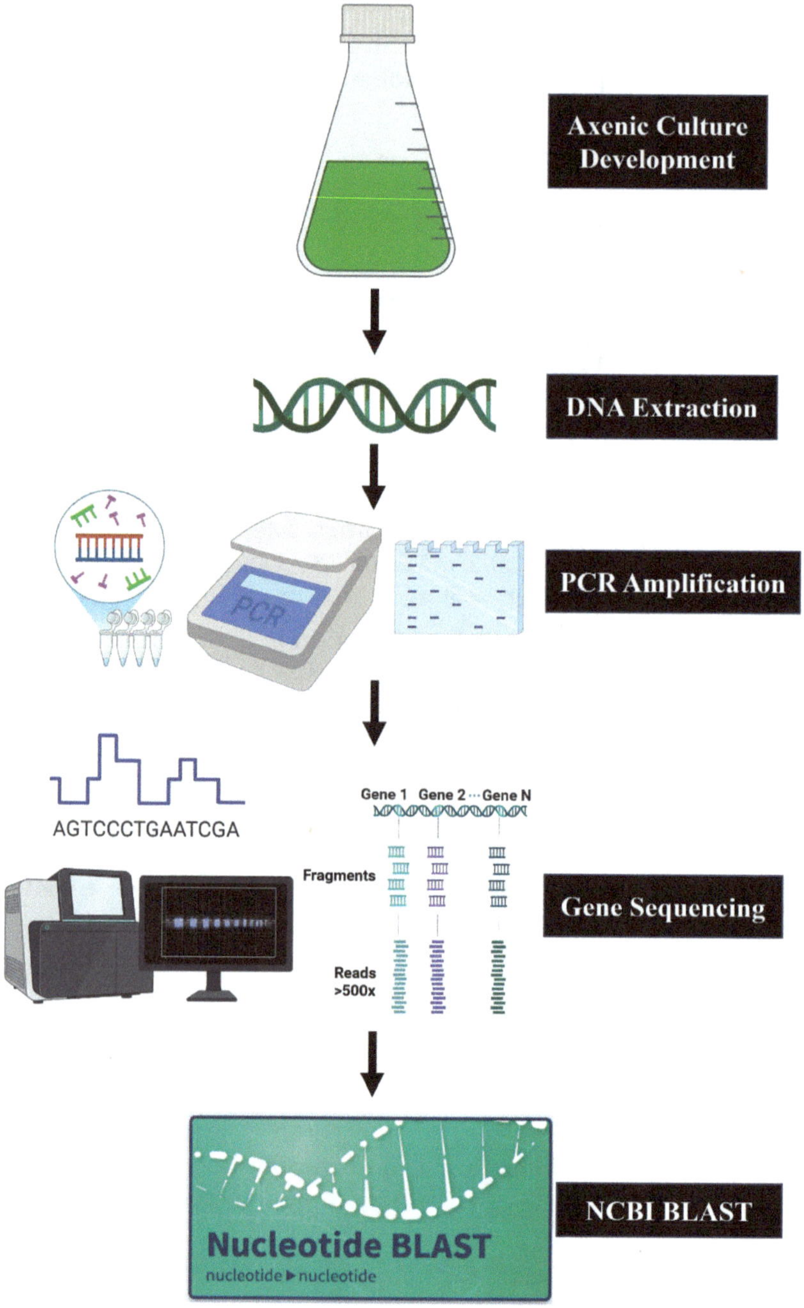

Fig. 1 Overall schematic representation of *18S rRNA* Gene Sequencing and BLAST Analysis Using NCBI

3. Analyze the top BLAST hits by the percentage identity, alignment length, and E-value (prefer >97–99% of identity with high coverage) [11] (Fig. 1).

4 Notes

1. Before DNA extraction, ensure the *Chlorella* isolate is monoclonal and axenic.
2. Assess both the quality and quantity of the extracted DNA to validate the high molecular weight DNA and absence of degradation.
3. Maintain the PCR amplicon length in suitable size (usually around 800–1000 bp and preferably ~500–800 bp). Longer amplicons might generate incomplete overlap or poor-quality reads at the ends, resulting in complicated assembly.
4. Only high-quality reads should be used for assembly.
5. Once a high-quality sequence is obtained, deposit it with full annotation and data to a public repository which supports the scientific community and enriches the reference databases.

Acknowledgments

We sincerely acknowledge the financial support from the DBT (Govt. of India), for the establishment of National Repository for Microalgae and Cyanobacteria—Freshwater & Marine (NRMC-F & M) (Phase II) (BT/PR29901/PBD26/694/2018). We gratefully acknowledge the Chief Minister Research Grant (CMRG), Govt. of Tamil Nadu for providing financial support (CMRG 2023-2024 Project No. CMRG2023BMBO7047), Rashtriya Uchchatar Shiksha Abhiyan (RUSA)—2.0 Biological Sciences (TN RUSA: 311/RUSA (2.0)/2018 Dt. December 02, 2020) and School of Life Sciences, Bharathidasan University, Tiruchirappalli, Tamil Nadu, India.

References

1. Bradley IM, Pinto AJ, Guest JS (2016) Design and evaluation of Illumina MiSeq-compatible, 18S rRNA gene-specific primers for improved characterization of mixed phototrophic communities. Appl Environ Microbiol 82(19): 5878–5891
2. Uddin MS, Cheng Q (2015) Recent application of biotechniques for the improvement of mango research. In: Applied plant genomics and biotechnology. Woodhead Publishing, pp 195–212
3. Hadziavdic K, Lekang K, Lanzen A, Jonassen I, Thompson EM, Troedsson C (2014) Characterization of the 18S rRNA gene for designing universal eukaryote specific primers. PLoS One 9(2):e87624
4. Crossley BM, Bai J, Glaser A, Maes R, Porter E, Killian ML, Clement T, Toohey-Kurth K (2020) Guidelines for Sanger sequencing and molecular assay monitoring. J Vet Diagn Invest 32(6):767–775
5. Johnson M, Zaretskaya I, Raytselis Y, Merezhuk Y, McGinnis S, Madden TL (2008) NCBI BLAST: a better web interface. Nucleic Acids Res 36(suppl_2):W5–W9
6. Haddad R, Alemzadeh E, Ahmadi AR, Hosseini R, Moezzi M (2014) Identification of Chlorophyceae based on 18S rDNA

sequences from Persian gulf. Iranian J Microbiol 6(6):437

7. Hossen R, Courtney M, Sim A, Khan MAAK, Verbruggen H, Bringloe T (2025) An optimized CTAB method for genomic DNA extraction from green seaweeds (Ulvophyceae). Applications in Plant Sciences 13(1): e11625
8. Jagielski T, Gawor J, Bakuła Z, Zuchniewicz K, Żak I, Gromadka R (2017) An optimized method for high quality DNA extraction from microalga Prototheca wickerhamii for genome sequencing. Plant Methods 13(1):77
9. Kim BH, Ramanan R, Cho DH, Choi GG, La HJ, Ahn CY, Oh HM, Kim HS (2012) Simple, rapid and cost-effective method for high quality nucleic acids extraction from different strains of Botryococcus braunii. PLoS One 7(5):e37770
10. Lanzén A, Lekang K, Jonassen I, Thompson EM, Troedsson C (2016) High-throughput metabarcoding of eukaryotic diversity for environmental monitoring of offshore oil-drilling activities. Mol Ecol 25(17):4392–4406
11. Attwood TK (2018) A Critical Guide to BLAST. F1000Res 7(1435):1435

Chapter 12

Phylogenetic Analysis of *18S rRNA* Gene in *Chlorella* Using Neighbor-Joining (NJ) Tree

Guhanraj Radhamanalan and Dhanasekaran Dharumadurai

Abstract

The genetic relationships among *Chlorella* species remain important for understanding their diversity and potential applications in environmental and industrial fields. In the present work, sequences collected from NCBI, selected *Chlorella* isolates were examined using MEGA11 to clarify their evolutionary relatedness. The sequences were aligned, the most suitable substitution model was identified, and a phylogenetic tree was generated using a Neighbor Joining (NJ) tree approach. Bootstrap testing was carried out to verify the stability of the branches. This analysis delivers a clear view of the *Chlorella* species are related and confirms the usefulness of MEGA11 for molecular comparison and species differentiation.

Key words *Chlorella* spp., MEGA 11, Neighbor Joining (NJ) tree, Molecular evaluation

1 Introduction

Chlorella is a genus of unicellular green microalgae found in freshwater, marine, and terrestrial environments [1, 2]. Although these species appear similar under the microscope, they show considerable genetic variation, making molecular markers essential for accurate identification. The *18S rRNA* gene is widely used for this purpose because conserved and variable regions allow reliable comparison among closely related taxa [3, 4]. Phylogenetic analysis of *18S rRNA* sequences helps determine the taxonomic position of isolates, assess their relationships with reference strains, and reveal diversity within a group of samples. This protocol reveals a concise workflow for analyzing *18S rRNA* sequences of *Chlorella* spp., including sequence retrieval, alignment, model selection, and phylogenetic tree construction using MEGA11.

Dhanasekaran Dharumadurai (ed.), *Chlorella Genomics, Cultivation, and Bioenterprise: Green Economy through Algal Innovations*, Springer Protocols Handbooks, https://doi.org/10.1007/978-1-0716-5276-3_12,
© The Author(s), under exclusive license to Springer Science+Business Media, LLC, part of Springer Nature 2026

2 Materials

1. Collected all *18S rRNA* sequences from NCBI (your isolates + reference *Chlorella* spp).
2. *Chlorella* sequence data file (FASTA format).
3. MEGA11 software.
4. NCBI GenBank database.
5. Computer or laptop with internet access.
6. ClustalW/Muscle alignment tools.

3 Methods

1. *Collection and Preparation of Sequences*
   (a) Open the NCBI Nucleotide database and download the *18S rRNA* gene sequences for the isolates and approximately 15 reference *Chlorella* species https://www.ncbi.nlm.nih.gov/nuccore/PX498621.1?report=fasta.
   (b) Combine all downloaded sequences into a single FASTA file.
   (c) Edit the sequence headers to keep only the accession number and culture name, removing unnecessary description.
   Example:
   >PX550085.1 *Chlorella sorokiniana.*
   >PX499712.1 *Chlorella sorokiniana isolate 21X1*
2. *Import of Sequences into MEGA 11*
   (d) Install MEGA 11 software and open the program.
   (e) Select Align → Edit/Build Alignment and import the FASTA file using Insert Sequences from File [5–7].
3. *Multiple Sequence Alignment*
   (f) Open Align in Alignment Explorer and run the alignment using MUSCLE or ClustalW.
4. *Trimming of Alignment*
   (g) Go to Alignment → Filter/Mask Sites.
   (h) Remove columns containing more than 20% gaps and exclude highly variable regions.
   (i) Save the trimmed alignment in MEGA format (.meg) using Data → Export Alignment.
5. *Selection of Substitution Model*
   (j) In the main MEGA window, choose Models → Find Best DNA/Protein Models (ML).
   (k) Load the trimmed alignment to identify the most suitable nucleotide substitution model (Fig. 1).

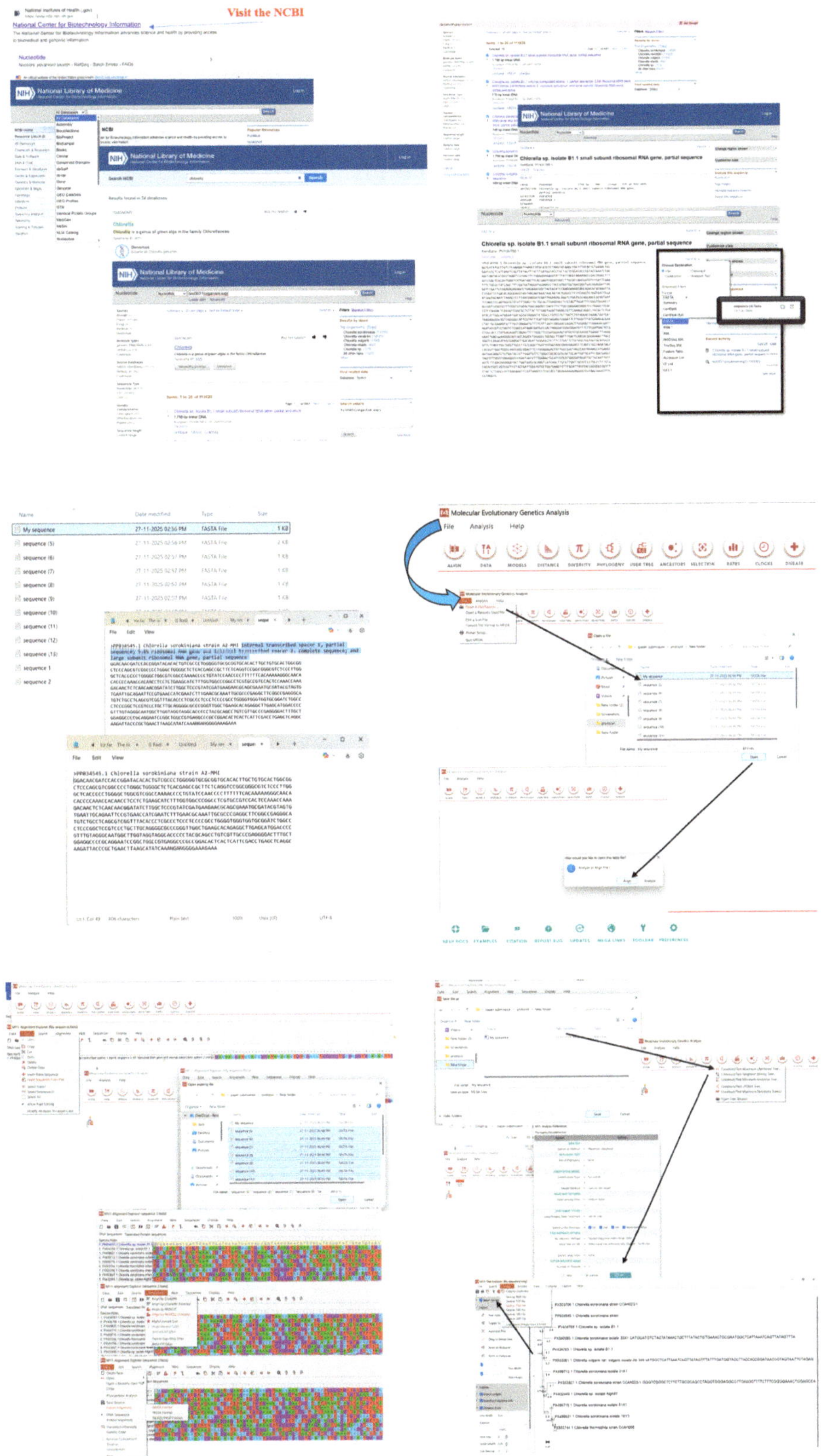

Fig. 1 Stepwise Workflow for phylogenetic tree construction using MEGA11

6. *Construction of Neighbor-Joining Tree*
   (l) Open Phylogeny → Construct Neighbor-Joining Tree.
   (m) Apply the model recommended by MEGA.
   (n) Set the bootstrap replication number to 1000.
7. *Rooting and Annotation*
   (o) In Tree explorer, root the tree using an appropriate outgroup taxon by selecting Tree → Root → Root at Selected Node.
   (p) Display bootstrap values and adjust visual settings such as color and line width.
   (q) Annotate isolates or highlight major *Chlorella* clades.
8. *Exporting the Final Tree*
   (r) Save the completed phylogenetic tree as PNG, PDF, or SVG using the Image → Save As option (Fig. 2).

4 Notes

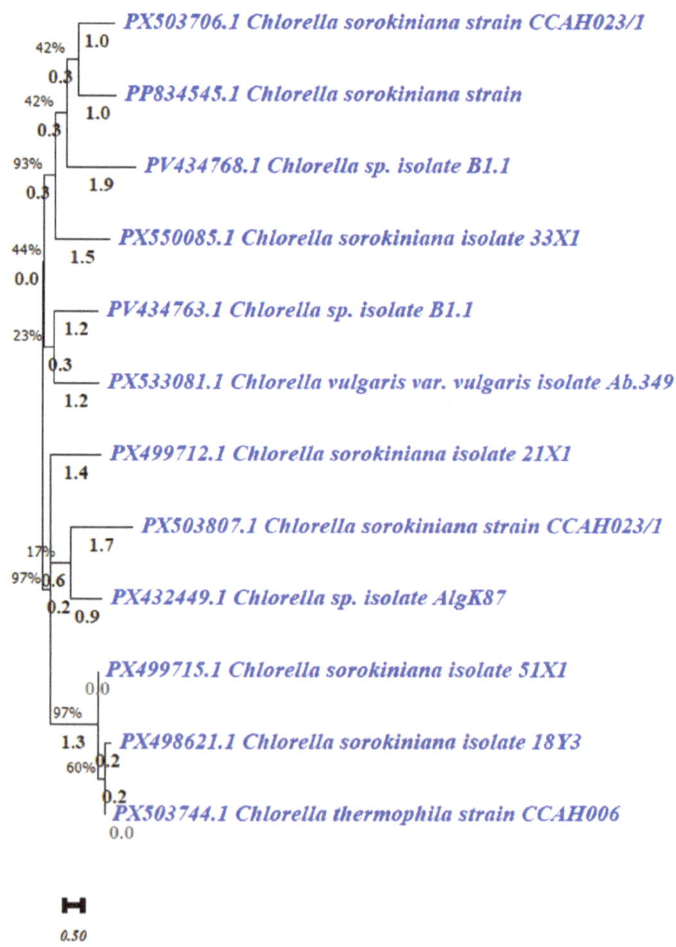

Fig. 2 Phylogenetic analysis of *Chlorella* spp. using MEGA11

References

1. Wu HL, Hseu RS, Lin LP (2001) Identification of *Chlorella* spp. isolates using ribosomal DNA sequences. Bot Bull Acad Sinica 42
2. Rajendran PJ, Shunmugam S, Saravanan C, Nooruddin T, Dharumadurai D (2025) Molecular monitoring, identification of the microcystin Synthetase (mcy) gene and Toxic potential of Microcystis aeruginosa in Indian freshwater blooms. Toxicon:108631
3. Goemann CL, Wilkinson R, Henriques W, Bui H, Goemann HM, Carlson RP et al (2023) Genome sequence, phylogenetic analysis, and structure-based annotation reveal metabolic potential of *Chlorella* sp. SLA-04. Algal Res 69: 102943
4. Chitthan SV, Dhanasekaran D, Thajuddin N (2025) Algal sensitivity and tolerance as key tools for validating organic pollution status in freshwater, detection of algal bioindicators. J Hydrol 661:133843
5. Smith MR (2022) Robust analysis of phylogenetic tree space. Syst Biol 71(5):1255–1270
6. Radhamanalan G, Dharumadurai D (2024) Anticandidal efficacy of oral probiont *Limosilactobacillus fermentum* against dental caries pathogens in children, Tamil Nadu, India. Indian J Microbiol 64(1):244–253
7. Comeau AM, Vincent WF, Bernier L, Lovejoy C (2016) Novel chytrid lineages dominate fungal sequences in diverse marine and freshwater habitats. Sci Rep 6(1):30120

Part III

Whole Genome Analysis of *Chlorella*

Chapter 13

Whole Genome Sequencing of *Chlorella* Using Illumina HiSeq

Guhanraj Radhamanalan

Abstract

Chlorella species are single-celled green microalgae of great importance in both environmental and biotechnological studies; yet, standardized procedures of whole genome sequencing are still scarce. The technical difficulties sequencing the genome of *Chlorella* include rigidity of cell walls and variation in DNA quality which influences the efficiency of sequencing and assembly results. This protocol details to complete a reproducible workflow of whole genome sequencing *Chlorella* with the Illumina HiSeq platform. Illumina HiSeq platform offers high sequencing depth and accuracy, which allows strong draft combined with assemblies of genomes. The protocol is flexible to various species of *Chlorella* and in the laboratory and aids in the production of high-quality genomic data of the microalgae.

Key words *Chlorella*, Illumina HiSeq, DNA library preparation, Genome assembly

1 Introduction

Chlorella species are common unicellular green microalgae that are in environmental, nutritional, and biotechnological study. They can be used in carbon sequestration and treatment of wastewater, biodiesel generation, and in the development of functional foods [1–3]. Nevertheless, the absence of standardized, reproducible whole genome sequencing methods has constrained the production of high-quality genomic resources of most *Chlorella* strains, especially with regard to the problems of their rigid cell walls and intricate eukaryotic genomes.

Whole genome sequencing is ideal in the prediction and functional annotation of genes along with comparison of genomes as the information is comprehensive and thus precise. Genome-scale data in *Chlorella* make it possible to discover metabolic pathways of photosynthesis, lipid storage, and adaptation to stress to improve the strain and conduct systems-level research [4, 5]. Although there have been improvements in the technology of sequencing, variation

Dhanasekaran Dharumadurai (ed.), *Chlorella Genomics, Cultivation, and Bioenterprise: Green Economy through Algal Innovations*, Springer Protocols Handbooks, https://doi.org/10.1007/978-1-0716-5276-3_13,
© The Author(s), under exclusive license to Springer Science+Business Media, LLC, part of Springer Nature 2026

in the efficiency of DNA extraction, the quality of library preparation, and in the data processing pipelines, still influences the results of genome assembly in microalgal studies. Illumina HiSeq platform provides high-throughput and high-accuracy and reproducible short-read sequencing, it is suitable to the draft genome sequencing of eukaryotic microalgae. Paired-end sequencing assists in effective assembly of genome in the presence of optimized laboratory and bioinformatics processes [6–8].

The present protocol is a workflow of sequencing entire genome of *Chlorella* under the Illumina HiSeq platform. The process includes algal growth, isolation of genomic DNA, preparation of the library, sequencing, and the downstream bioinformatics analysis. The procedure should be flexible to other *Chlorella* species and laboratory conditions to offer a convenient system to investigators who want to acquire quality-based genomic data on microalgal research.

2 Materials

1. Fresh water sample.
2. *Chlorella* sp. culture.
3. BG-11 growth medium.
4. Phosphate-buffered saline (PBS) pH = 7.4.
5. Sterile distilled water.
6. Sterile culture flasks.
7. Microcentrifuge tubes (1.5 mL and 2.0 mL).
8. Centrifuge tubes (15 mL and 50 mL).
9. Filter tips .
10. Mechanical lysis beads of glass.
11. Tris–HCl (pH 8.0).
12. EDTA (pH 8.0).
13. Sodium chloride (NaCl).
14. β-Mercaptoethanol.
15. Proteinase K.
16. RNase A.
17. Chloroform:isoamyl alcohol (24:1).
18. Isopropanol.
19. Ethanol (70%).
20. TE buffer/nuclease-free water.
21. Illumina-compatible kit of DNA library preparation.

22. Library purification using magnetic beads.
23. High-fidelity PCR master mix.
24. Illumina HiSeq sequencing reagent and flow cell.
25. Light controlled growth chamber.
26. Microcentrifuge.
27. Homogenizer.
28. Water bath.
29. Qubit 3 fluorometer.
30. Agarose gel electrophoresis system.
31. Fragment analyzer.
32. Magnetic stand.
33. FastQC.
34. Trimmomatic.
35. SPAdes.
36. QUAST.
37. BUSCO.
38. BLAST.

3 Methods

1. Sample collection
   - Fresh water sample was collected from Chennai region, Tamil Nadu, India.
2. Culture of *Chlorella*
   - *Chlorella* is cultured in BG-11 medium.
   - The cultures are kept at 25 °C with constant lighting.
   - Light aeration is accessible to make the growth homogeneous.
   - The harvesting of cells takes place in the period of exponential growth.
3. *Biomass harvesting*
   - Centrifugation of cultures is done at 5000 × g in 15 min.
   - Cell pellet is washed twice using sterile phosphate-buffered saline.
   - The collected biomass is frozen in a −80 °C and then an extraction of DNA was performed.

4. *Genomic DNA extraction*
   - About 100 mg of frozen *Chlorella* biomass is resettled into a sterile microcentrifuge tube.
   - There is an addition of CTAB lysis buffer with Tris–HCl, EDTA, NaCl, and proteinase K.
   - Glass beads and bead beating are used in mechanical cell disruption.
   - The incubation of the lysate is done at 65 °C, 30 min.
   - Centrifugation is completed after the addition of chloroform:isoamyl alcohol (24:1).
   - Aqueous stage is carefully transferred in a fresh tube.
   - Cold isopropanol is used to precipitate the genomic DNA.
   - DNA pellet was washed with 70% ethanol, air dried, and resuspended in nuclease free water or TE buffer [3, 4].
5. *DNA quality and quantity determinacy*
   - Agarose gel electrophoresis is used to determine the integrity of DNA.
   - The purity of DNA is determined using the ratio of A 260/A 280 and A 260/A 230.
   - The level of DNA is ascertained by a Qubit 3 fluorometer.
   - Only good DNA samples are chosen to be used in library preparation.
6. *DNA fragmentation*
   - gDNA is broken in pieces with an average size of about 350 bp.
   - Out shear is achieved either through acoustic shearing or enzyme.
7. *Library preparation*
   - The pieces of the DNA are end-repair to form blunt fragments.
   - A base of adenine is inserted on 3 ends of the DNA fragments.
   - Unique indices of adapters used in illumina sequencing are joined to the fragments of DNA.
   - Short adapter-ligated DNA is amplified by limited-cycle PCR.
   - AMPure beads based on a particular cleanup method are used to clean PCR products (Fig. 1).

8. *Quality control of library*
   - The concentration of library can be measured by Qubit 3 fluorometry.
   - Normal and pooling of libraries that satisfy the quality standards occurs.
9. *Illumina HiSeq sequencing*
   - The libraries are pooled onto Illumina HiSeq flow cell.
   - The sequencing was used to perform paired-end sequencing (2×150 bp).
   - FASTQ format is used to produce raw sequencing data.
10. *Raw read quality control*
    - Read quality analysis software is used to assess sequencing reads in quality.
    - Adaptor sequences and poor-quality bases are DNA clipping out by using Trimmomatic.
    - Reads that fall below set quality bases are eliminated.
    - Good reads are stored to be used in genome assembly.
11. *Genome assembly*
    - The assemblies of filtered reads are assembled de novo with a genome assembly tool created to optimally work with eukaryotic microalgae (Fig. 2).
    - The quality of assemblies (QUAST) is measured as genome size, N_{50} value and GC content [5, 6, 8].
12. *Genome annotation*
    - The automated genome annotation pipelines are utilized to predict the genes.
    - Protein and pathway databases that are publicly available are used to complete the functional annotation process [8–10].
13. *Data storage and submission*
    - Raw and processed sequencing data are stored.
    - Finally, assemblies of genomes and raw read sequencing are uploaded to a public sequence repository (NCBI).

Flowchart

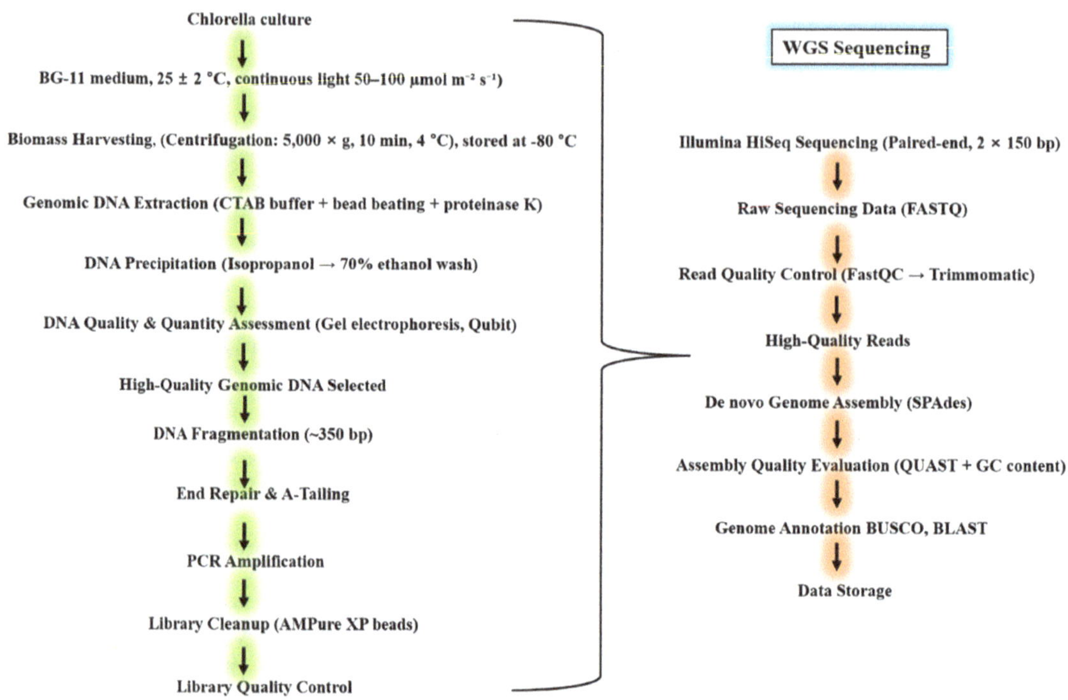

Fig. 1 Step-by-step workflow of whole-genome sequencing and bioinformatic analysis—*Chlorella* sp.

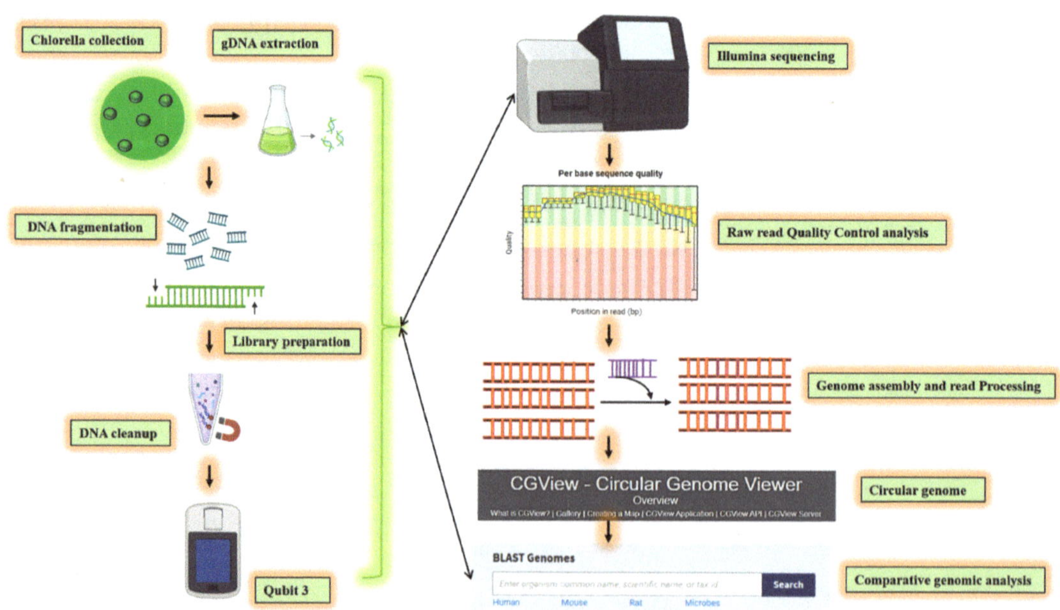

Fig. 2 Schematic workflow of whole-genome sequencing and bioinformatic analysis—*Chlorella* sp.

4 Notes

Parameter	Observation	Expected results
Algal culture growth		
Biomass yield		
Genomic DNA integrity		
DNA purity		
DNA concentration		
DNA fragmentation		
Library preparation		
Library size profile		
Library concentration		
Sequencing output		
Read quality		
Adapter contamination		
Genome assembly size		
Assembly continuity		
GC content		
Genome completeness		
Gene prediction		
Functional annotation		
Data submission		

References

1. Wu T, Li L, Jiang X, Yang Y, Song Y, Chen L et al (2019) Sequencing and comparative analysis of three Chlorella genomes provide insights into strain-specific adaptation to wastewater. Sci Rep 9(1):9514
2. Cecchin M, Marcolungo L, Rossato M, Girolomoni L, Cosentino E, Cuine S et al (2019) Chlorella vulgaris genome assembly and annotation reveals the molecular basis for metabolic acclimation to high light conditions. Plant J 100(6):1289–1305
3. Arolla RG, Naik KS (2024) Genomic characterization and comparative genomics of Chlorella sp. CH2018 from Musi River water, India. Gene Rep 37:102076
4. Radhamanalan G, Mickymaray S, Alothaim AS, Alsowayeh N, Dharumadurai D (2025) Whole-genome analysis of Streptococcus mutans GDP01 isolate of dental caries tooth. Microbiol Resour Announce 14(6):e01057–e01024
5. Wang B, Jia Y, Dang N, Yu J, Bush SJ, Gao S et al (2024) Near telomere-to-telomere genome assemblies of two Chlorella species unveil the composition and evolution of centromeres in green algae. BMC Genomics 25(1):356
6. Hovde BT, Hanschen ER, Tyler CRS, Lo CC, Kunde Y, Davenport K et al (2018) Genomic characterization reveals significant divergence

within Chlorella sorokiniana (Chlorellales, Trebouxiophyceae). Algal Res 35:449–461

7. Angelova A, Park SH, Kyndt J, Fitzsimmons K, Brown JK (2014) Sonication-based isolation and enrichment of Chlorella protothecoides chloroplasts for Illumina genome sequencing. J Appl Phycol 26(1):209–218

8. Guhanraj R, Dhanasekaran D (2024) Probiotic functional gene explorations in the genome of Limosilactobacillus fermentum GD5MG. Microb Pathog 192:106686

9. Khizar Y, Farooq U, Attia KA, Rehman OU, Abushady AM, Fiaz S et al (2025) Genome-wide identification and characterization of stress-responsive genes in Chlorella vulgaris. BMC Genomic Data 26(1):20

10. Goemann CL, Wilkinson R, Henriques W, Bui H, Goemann HM, Carlson RP et al (2023) Genome sequence, phylogenetic analysis, and structure-based annotation reveal metabolic potential of Chlorella sp. SLA-04. Algal Res 69:102943

Chapter 14

Chlorella Genome Assembly with SPAdes and Annotation Using GenSAS

Asraf Sithikka Rasheed, Shakena Fathima Thajuddin, Thajuddin Nooruddin, and Dhanasekaran Dharumadurai

Abstract

Generation of high-quality genome assemblies and their subsequent annotation. In order to anticipate both the structural and functional features of genes, genome assembly is a crucial step to proceed further genomic analysis. Genome annotation processes integrate alignment-based evidence with ab initio gene predictions to generate an absolute collection of gene annotations. Even though eukaryotic genome annotation is now more feasible for researchers without specialized expertise, it is still very difficult for researchers who are not expert in bioinformatics or computational biology. Genome annotation converts the raw sequence data into a functional form by identifying the key aspects of the genes and their products. As the annotation tools are rapidly developing, the field of genomic research is augmenting over this high-quality information to drive all areas of biological research.

Key words *Chlorella* genome, Genome assembly, SPAdes, Genome annotation, GenSAS

1 Introduction

SPAdes is a bacterial genome assembly algorithm based on de Bruijn graph published in 2012 [1]. The St. Petersburg genome Assembler (SPAdes) was initially created for de novo assembly of small genomes, such as those from prokaryotes and single cells. But now, its scope has been significantly expanded to encompass metagenomes, transcriptomes, and plasmids. SPAdes incorporates various sequencing technologies, rendering robust hybrid assembly pipelines for integrating the accuracy of short reads (Illumina) with the continuity of long reads (PacBio & Oxford Nanopore) [2–4]. SPAdes employ specialized de Bruijn graph algorithms enhanced by specialized techniques such as k-bimer adjustment and paired assembly graphs [3, 8]. SPAdes excel with complex datasets, including those from single cell or metagenomic sequencing, which often suffer from uneven coverage and high error rates.

Dhanasekaran Dharumadurai (ed.), *Chlorella Genomics, Cultivation, and Bioenterprise: Green Economy through Algal Innovations*, Springer Protocols Handbooks, https://doi.org/10.1007/978-1-0716-5276-3_14,
© The Author(s), under exclusive license to Springer Science+Business Media, LLC, part of Springer Nature 2026

It consistently outperforms leading assemblers like Velvet, IDBA-UD, and SOAP de novo, by generating longer, more accurate contigs with superior N50 scores, particularly in challenging, low coverage samples [3, 5–8]. While even the large genomes may now be sequenced in a single lab due to the low cost of sequencing in recent days, the subsequent task of genome annotation has become paradoxically more challenging. Compared to earlier massive attempts, modern genome annotation projects are typically smaller in scale and are frequently carried out by researchers with limited expertise in computational biology. Genome annotation processes integrate alignment-based evidence with ab initio gene predictions to generate an absolute collection of gene annotations [9]. Genome Sequence Annotation Server (GenSAS) is a comprehensive, web-based genome annotation tool. Multiple structural and functional annotation tools have been consolidated into one user-friendly interface in this integrated web platform. The software offers an efficient interface with graphics for all phases of the process, from manual curation to analysis and visualization [10, 11]. Key annotation processes, including as repeat masking, evidence alignment, structural and functional annotation, and final file output, which can also be edited manually, are guided by the GenSAS workflow. Moreover, the flexibility of this system allows for the continuous addition of new tools and advancement of its accessible resources [12]. This chapter provides a comprehensive protocol for the eukaryotic genome assembly and annotation using SPAdes and GenSAS.

2 Materials

1. Whole Genome Sequence (WGS) data.
2. ART Illumina NGS Read Simulator software.
3. FastQC tool, Trimmomatic or Cutadapt tool.
4. SPAdes Genome Assembler v3.9.0.
5. QUAST.
6. GenSAS web server.
7. JBrowse/Apollo tools.

3 Methodology

1. If a completed genome of *Chlorella* sp. is available for a particular strain, you can simulate reads for optional validation in order to test and optimize assembly configurations before utilizing actual data.

3.1 Genome Assembly Using SPAdes

3.1.1 Illumina Read Simulation of Genomes

2. The reference genome sequence is first obtained from a source such as NCBI GenBank, and then the ART Illumina read simulator is used to simulate Illumina paired-end reads.
3. The simulation should be set up for a library with 150 bp read lengths, a 450 bp insert size, a 50 bp standard deviation, and multiple coverages of 100x, 50x, and 25x to assess performance across various depths. It can also employ an error profile derived from your particular sequencing platform (such as HiSeq 2000 or NovaSeq 6000), or a built-in profile if a custom one is not available [13, 14].

3.1.2 Data Preparation

1. For a true de novo assembly, obtain Whole Genome Sequencing (WGS) data for *Chlorella* sp. from a short-insert, paired-end Illumina library.
2. Run FastQC on the raw FASTQ files to evaluate initial read quality and identify potential issues such as adapter contamination or low-quality bases.
3. Process the raw reads using a tool like Trimmomatic or Cutadapt to remove adapter sequences, trim low-quality bases from the start and end of reads, discard any reads that fall below a minimum length threshold after trimming.
4. Re-run FastQC on the trimmed and filtered FASTQ files to confirm successful adapter removal and an overall improvement in read quality before proceeding to assembly [14].

3.1.3 Genome Assembly

1. Install the required genome assembler—SPAdes 3.9.0, ensuring version control for reproducibility.
2. Run each installed assembler using a defined range of k-mer values.
3. For each assembler, at least 5 k-mers (21, 31, 51, 75, and 89) were applied, and the best assembly (based on NGA50) was used as a representative of each assembler for comparison [14].
4. Evaluate the quality of each assembly by comparing it to a reference genome (if available) or using intrinsic metrics [14, 15].

3.2 Genome Annotation Using GenSAS

1. To begin using GenSAS, log in on the GenSAS homepage (https://www.gensas.org/).
2. Open the main GenSAS annotation interface by clicking the "Use GenSAS" tab (Fig. 1).
3. Section A used to navigate the workflow using the header flowchart. The horizontal flowchart shows the complete annotation process. Each color of the arrow indicates the status of annotation steps Gray signifies the unavailability of the step,

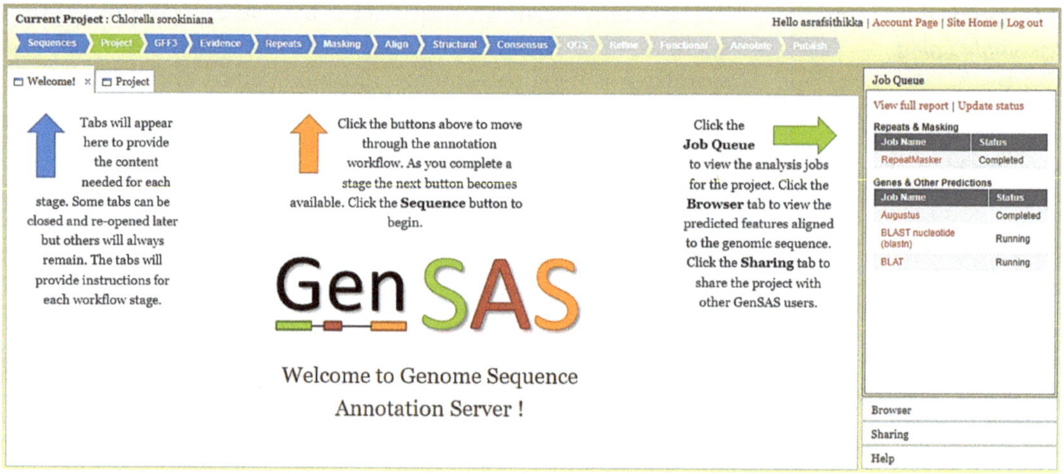

Fig. 1 GenSAS main annotation tab

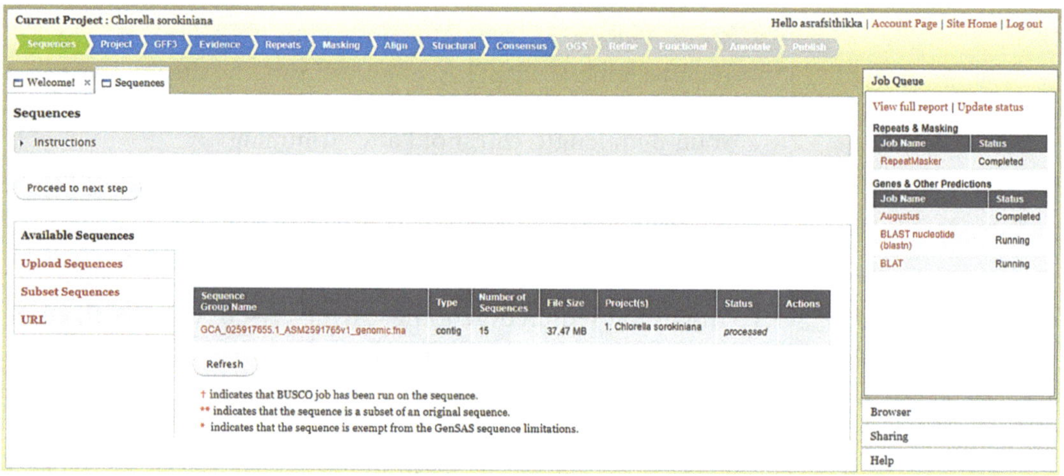

Fig. 2 Uploading genome sequence for annotation

Blue depicts the available step, and Green shows the currently viewing step.

4. Section B is the primary workspace where the steps are performed such as uploading files, configuring settings, and submitting jobs for each step (Figs. 2, 3, and 4).
5. Section C provides rapid access to essential tools. Additionally, JBrowse/Apollo can be used to visualize the genome annotations (Fig. 5).
6. Monitor and update the status of your submitted jobs using the job queue.

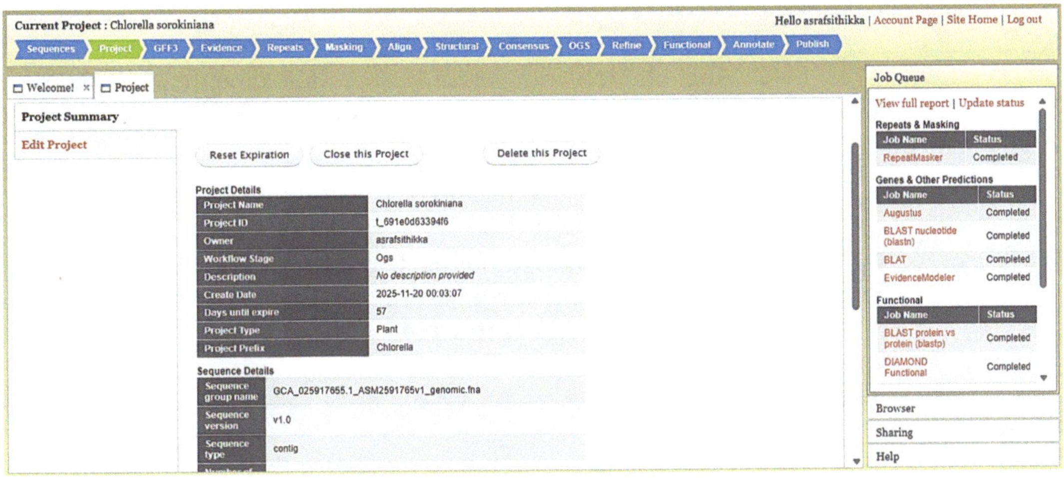

Fig. 3 Project Summary of the uploaded genome

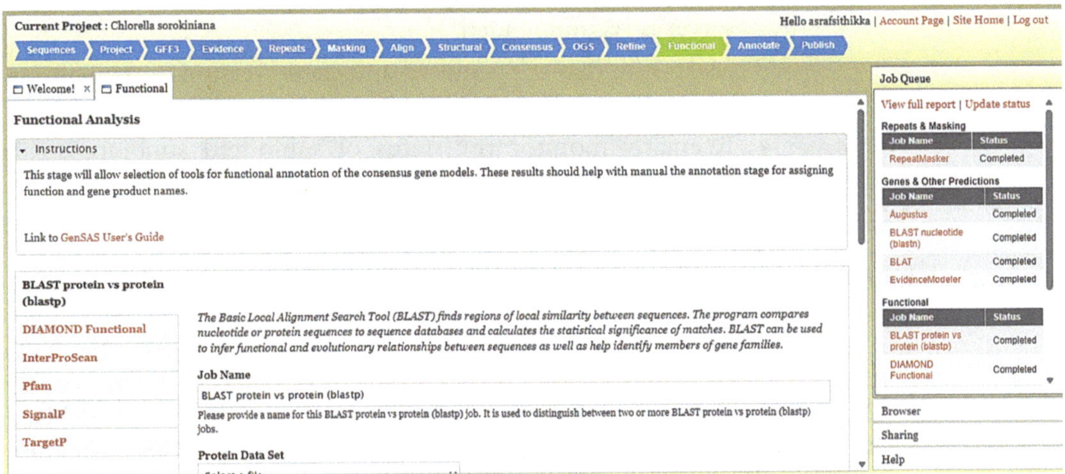

Fig. 4 Accession of various tools for genome annotation

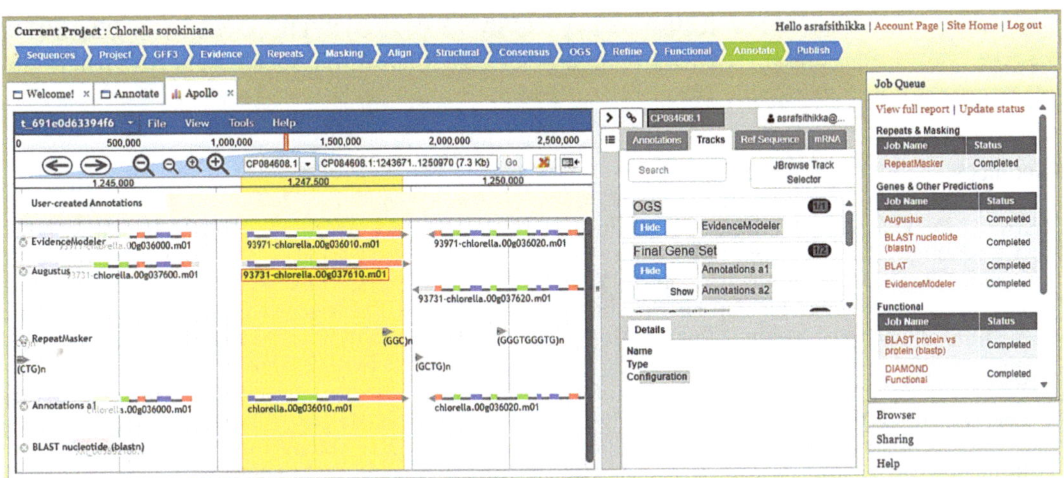

Fig. 5 Visualization of genome annotation using Apollo platform

7. After the job completion, view results directly in JBrowse, or access detailed outputs, including GFF3 files and summary tables through the job queue.
8. To optimize the parameters, submit multiple simultaneous tasks for the same tool.
9. Inspect the annotations in JBrowse regularly and continuously refine the results by adjusting the parameters of the jobs and resubmitting (*see* **Notes 1–4**).

4 Notes

1. Ensure the adequate memory (sufficient RAM) and processing time provided throughout the experiment.
2. High-performance computing (HPC) system like Linux-based server is recommended.
3. Make sure that the standard and stable version of the software and web-server are employed.
4. Regularly monitor the status of submitted annotation jobs through the job queue in the user interface of GenSAS.

Acknowledgments

We sincerely acknowledge the financial support from the DBT (Govt. of India), for the establishment of National Repository for Microalgae and Cyanobacteria—Freshwater & Marine (NRMC-F & M) (Phase II) (BT/PR29901/PBD26/694/2018). We gratefully acknowledge the Chief Minister Research Grant (CMRG), Govt. of Tamil Nadu for providing financial support (CMRG 2023-2024 Project No. CMRG2023BMBO7047), Rashtriya Uchchatar Shiksha Abhiyan (RUSA) – 2.0 Biological Sciences (TN RUSA: 311/RUSA (2.0)/2018 Dt. December 02, 2020) and School of Life Sciences, Bharathidasan University, Tiruchirappalli, Tamil Nadu, India.

References

1. Merda D, Vila-Nova M, Bonis M, Boutigny AL, Brauge T, Cavaiuolo M, Cunty A, Regnier A, Sayeb M, Vingadassalon N, Yvon C (2024) Unraveling the impact of genome assembly on bacterial typing: a one health perspective. BMC Genomics 25(1):1059
2. Prjibelski A, Antipov D, Meleshko D, Lapidus A, Korobeynikov A (2020) Using SPAdes de novo assembler. Curr Protoc Bioinformatics 70(1):e102
3. Bankevich A, Nurk S, Antipov D, Gurevich AA, Dvorkin M, Kulikov AS, Lesin VM, Nikolenko SI, Pham S, Prjibelski AD, Pyshkin AV (2012) SPAdes: a new genome assembly algorithm and its applications to single-cell sequencing. J Comput Biol 19(5):455–477
4. Antipov D, Korobeynikov A, McLean JS, Pevzner PA (2016) hybridSPAdes: an algorithm for hybrid assembly of short and long reads. Bioinformatics 32(7):1009–1015

5. Çabuk U, Ünlü ES (2022) A combined de novo assembly approach increases the quality of prokaryotic draft genomes. Folia Microbiol 67(5):801–810
6. Papudeshi B, Haggerty JM, Doane M, Morris MM, Walsh K, Beattie DT, Pande D, Zaeri P, Silva GG, Thompson F, Edwards RA (2017) Optimizing and evaluating the reconstruction of metagenome-assembled microbial genomes. BMC Genomics 18(1):915
7. Tizabi D, Bachvaroff T, Hill RT (2022) Comparative analysis of assembly algorithms to optimize biosynthetic gene cluster identification in novel marine actinomycete genomes. Front Mar Sci 9:914197
8. Nurk S, Bankevich A, Antipov D, Gurevich AA, Korobeynikov A, Lapidus A, Prjibelski AD, Pyshkin A, Sirotkin A, Sirotkin Y, Stepanauskas R (2013) Assembling single-cell genomes and mini-metagenomes from chimeric MDA products. J Comput Biol 20(10):714–737
9. Yandell M, Ence D (2012) A beginner's guide to eukaryotic genome annotation. Nat Rev Genet 13(5):329–342
10. Lee T, Peace C, Jung S, Zheng P, Main D, Cho I (2011, October) GenSAS—an online integrated genome sequence annotation pipeline. In: 2011 4th international conference on biomedical engineering and informatics (BMEI), vol 4. IEEE, pp 1967–1973
11. Ejigu GF, Jung J (2020) Review on the computational genome annotation of sequences obtained by next-generation sequencing. Biology 9(9):295
12. Humann JL, Lee T, Ficklin S, Main D (2019) Structural and functional annotation of eukaryotic genomes with GenSAS. In: Gene prediction: methods and protocols. Springer New York, New York, NY, pp 29–51
13. Huang W, Li L, Myers JR, Marth GT (2012) ART: a next-generation sequencing read simulator. Bioinformatics 28:593–594
14. Forouzan E, Maleki MSM, Karkhane AA, Yakhchali B (2017) Evaluation of nine popular de novo assemblers in microbial genome assembly. J Microbiol Methods 143:32–37
15. Gurevich A, Saveliev V, Vyahhi N, Tesler G (2013) QUAST: quality assessment tool for genome assemblies. Bioinformatics 29:1072–1075

Chapter 15

Genome Characterization of *Chlorella* Using GeSeq

Fayaazuddin Thajuddin, Thajuddin Nooruddin, and Dhanasekaran Dharumadurai

Abstract

Genome characterization is a critical initial step to understand the functional, structural, and evolutionary features of ecologically, biotechnologically, and economically important microalgae such as *Chlorella* spp., which are widely employed in wastewater remediation, biofuel production, development of various pharmaceutical and nutraceutical products, and diverse biomaterials. Because of the accessibility of high-quality whole genome and organelle genome sequences, in silico genome annotation platforms have become essential for rapid and precise genome analyses. GeSeq, a section of the CHLOROBOX webserver, is a specialized web-based tool that has been developed for accurate chloroplast and other organellar genome annotation by employing curated reference databases, BLAST-based homology searches, and integrated tRNA prediction algorithms such as tRNAscan-SE and ARAGORN. This protocol outlines a comprehensive GeSeq approach that includes genome analysis and visualization using GeSeq and OGDRAW tools.

Key words *Chlorella*, Genome characterization, CHLOROBOX, GeSeq, OGDRAW

1 Introduction

Whole genome analysis of *Chlorella* has significantly advanced with the accessibility of the high-quality genome sequences. In order to develop an efficient, standard, and cost-effective algal cultivation system, an extensive insight is required on both the physiological and genetic characteristics of the strains. This understanding can be achieved through the genome sequencing and characterization, which enables the acceleration of the process of strain optimization and enhances the hypothesis-driven research [1, 2]. The emergence of various advancements in genomics, multi-omics, and synthetic biology allows for the highly precise genetic engineering in microalgae and cyanobacteria to acquire the desired characteristics to obtain enhanced yield of specific compounds. The developing of genome scale metabolic models is made easier by in-silico flux balance analysis by the employment of genomic and proteomics data, which significantly lowers the requirement of extensive and

Dhanasekaran Dharumadurai (ed.), *Chlorella Genomics, Cultivation, and Bioenterprise: Green Economy through Algal Innovations*, Springer Protocols Handbooks, https://doi.org/10.1007/978-1-0716-5276-3_15,

© The Author(s), under exclusive license to Springer Science+Business Media, LLC, part of Springer Nature 2026

time-consuming wet lab studies [3]. GeSeq is an online tool designed for the rapid and precise annotation of the organellar genome with the specialized function for chloroplast genomes. GeSeq renders a manually curated library of reference sequences for chloroplast genomes and allows for the user-friendly approach to select or upload the appropriate reference sequences for each project. The primary output of GeSeq is a GenBank file and OGDRAW for instant visualization and typically requires minimal curation. Additionally, GeSeq provides a range of optional secondary outputs that support downstream analyses, such as phylogenetic or comparative genomic research [4]. This chapter provides a comprehensive step by step protocol for the analysis and characterization of organellar genome of *Chlorella* using GeSeq.

2 Materials

1. *Chlorella* genome (FASTA format).
2. Reference genome of closely related species (Optional).
3. GeSeq and OGDRAW tools under CHLOROBOX webserver.

3 Methodology

1. Go to the GeSeq tool (https://chlorobox.mpimp-golm.mpg.de/geseq.html) under the CHLOROBOX webserver (Fig. 1).
2. Upload the FASTA sequence of the *Chlorella* genome to be analysed (Fig. 2).

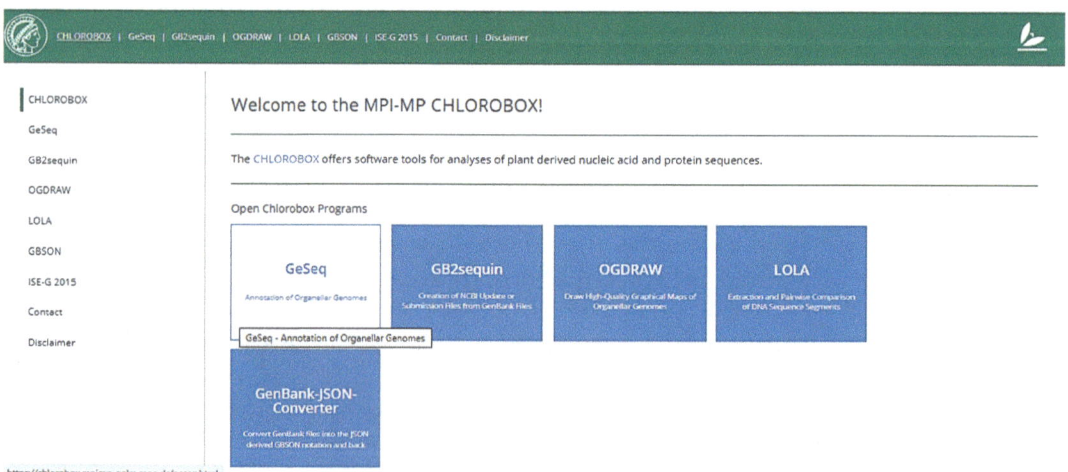

Fig. 1 Homepage of CHLOROBOX (Selecting GeSeq tool)

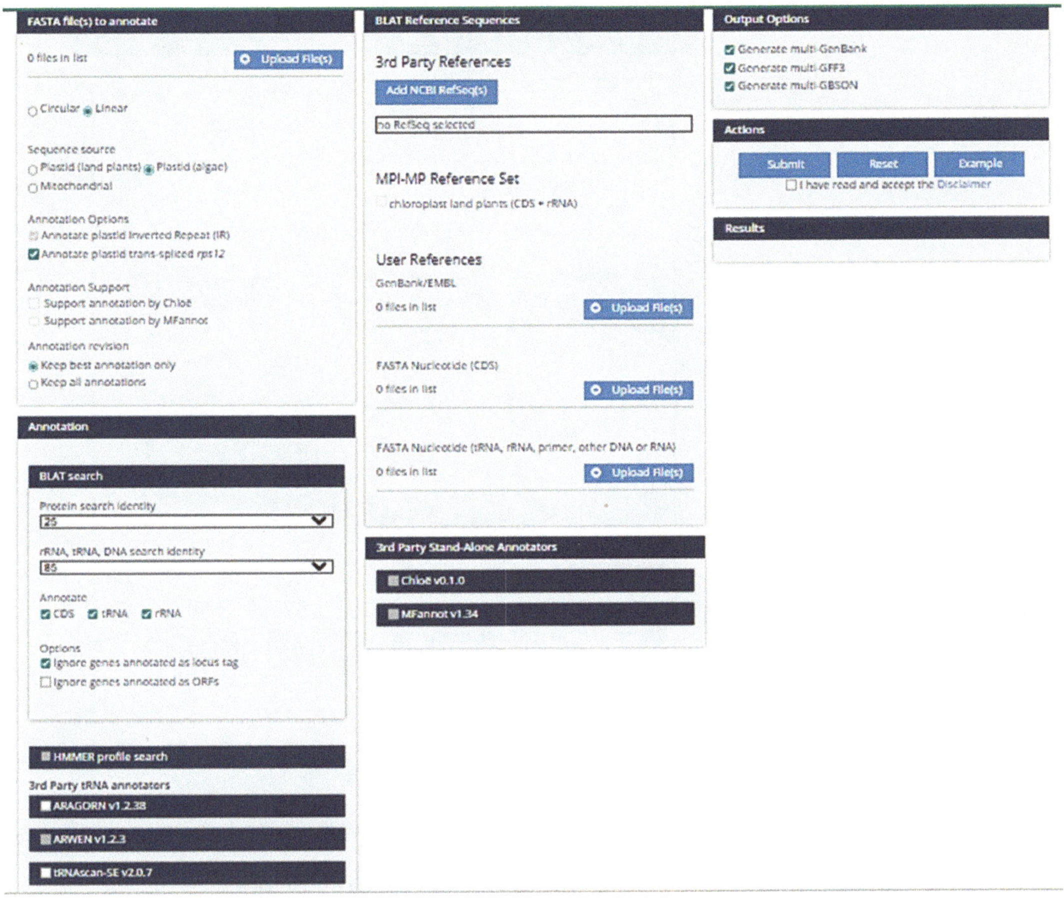

Fig. 2 GeSeq interface showing input options, reference selection, annotation tools, and output settings for organelle genome annotation

3. Upload a GenBank file from a closely related species directly in GeSeq as custom reference or select a reference from NCBI within the tool to facilitate the precise annotation (optional).
4. Enable BLAST (blastn/blastx) against the selected references.
5. Select both tRNAscan-SE and ARAGORN within GeSeq to annotate tRNAs. Using both tools is recommended for complementary coverage where tRNAscan–SE ensures high sensitivity, and ARAGORN also identifies tmRNA.
6. Specify identity/coverage thresholds like the BLAST identity of $\geq$80–90% for the initial results.
7. Click submit to run GeSeq and monitor the job status in the results section.
8. After running, the annotated features can be accessed in GenBank, GFF3, and the format of files suitable for the

Fig. 3 Result page of GeSeq submitted project

visualization through OGDRAW and the reports can also be obtained as summary tables (Fig. 3).

9. Run OGDRAW, another tool within the CHLOROBOX web-server to visualize the organelle genome map by employing the output from GeSeq project (Fig. 4) [4, 5].
10. Adjust the colors, labels, and gene order display as per desire and clarity (*see* **Notes 1–4**).

4 Notes

1. Ensure the adequate memory (sufficient RAM) and processing time provided throughout the experiment.
2. Ensure the stable and uninterrupted internet connection throughout the experiment.
3. Make sure that the standard and stable version of the web-server is employed.
4. Regularly monitor the status of submitted jobs through the job queue in the results section of the GeSeq page.

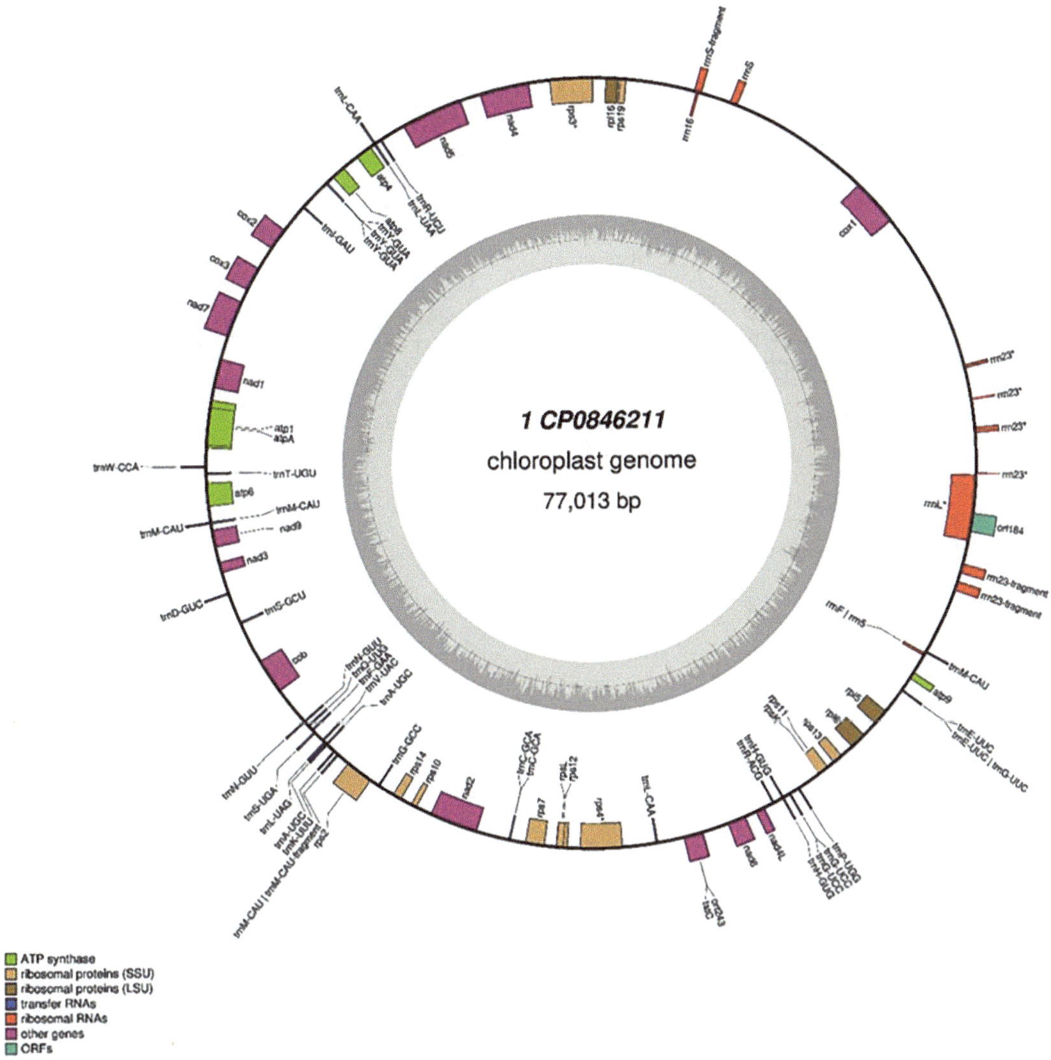

Fig. 4 Genome visualization using OGDRAW

Acknowledgments

We sincerely acknowledge the financial support from the DBT (Govt. of India), for the establishment of National Repository for Microalgae and Cyanobacteria—Freshwater & Marine (NRMC-F & M) (Phase II) (BT/PR29901/PBD26/694/2018). We gratefully acknowledge the Chief Minister Research Grant (CMRG), Govt. of Tamil Nadu for providing financial support (CMRG 2023-2024 Project No. CMRG2023BMBO7047), Rashtriya Uchchatar Shiksha Abhiyan (RUSA) – 2.0 Biological Sciences (TN RUSA: 311/RUSA (2.0)/2018 Dt. December 02, 2020) and Department of Microbiology, Bharathidasan University, Tiruchirappalli, Tamil Nadu, India.

References

1. Unkefer CJ, Sayre RT, Magnuson JK, Anderson DB, Baxter I, Blaby IK, Brown JK, Carleton M, Cattolico RA, Dale T, Devarenne TP (2017) Review of the algal biology program within the National Alliance for advanced biofuels and bioproducts. Algal Res 22:187–215
2. Hovde BT, Hanschen ER, Tyler CRS, Lo CC, Kunde Y, Davenport K, Daligault H, Msanne J, Canny S, Eyun SI, Riethoven JJM (2018) Genomic characterization reveals significant divergence within Chlorella sorokiniana (Chlorellales, Trebouxiophyceae). Algal Res 35: 449–461
3. Thajuddin F, Rasheed AS, Elumalai A, Palanivel P, Nooruddin T, Dharumadurai D (2026) Harnessing microalgae and cyanobacteria for sustainable bioenergy production: integration of molecular engineering and AI-driven approaches. Renew Energy 256:123961
4. Tillich M, Lehwark P, Pellizzer T, Ulbricht-Jones ES, Fischer A, Bock R, Greiner S (2017) GeSeq–versatile and accurate annotation of organelle genomes. Nucleic Acids Res 45(W1): W6–W11
5. Greiner S, Lehwark P, Bock R (2019) OrganellarGenomeDRAW (OGDRAW) version 1.3.1: expanded toolkit for the graphical visualization of organellar genomes. Nucleic Acids Res 47: W59–W64

Chapter 16

Metabolic Pathway Analysis of the *Chlorella* Genome Using EggNOG-Mapper and KEGG

Ramakrishnan Govindharaj Kumar, Gowtham Neethimohan, Dhinesh Kumar, and Dhanasekaran Dharumadurai

Abstract

Metabolic pathway analysis (MPA) provides researchers with a comprehensive view of an organism's genome, as described by its genes, and its biochemical capabilities and potential. A detailed MPA of organisms like *Chlorella* microalgae will help researchers investigate and develop biofuels and nutraceuticals, recover and recycle nutrients from wastewater, and maintain environmental sustainability. Although many new high-quality genome sequences are being generated, these are primarily unannotated at this time, so researchers must rely on in-silico functional annotation to create accurate metabolic pathways. EggNOG-mapper is a well-established and reliable annotation service that uses orthology-based comparison methods for classifying functional gene products. Kyoto Encyclopedia of Genes and Genomes (KEGG) allows researchers to reconstruct metabolic networks (using KO assignments) systematically. This chapter contains a comprehensive protocol describing step-by-step methods of using EggNOG-mapper and KEGG tools to perform annotation and metabolic pathway development of the *Chlorella* genome, along with a comprehensive and substantial understanding of key metabolic processes that potentially will provide additional biotechnological applications.

Key words *Chlorella* genome, EggNOG-mapper, KEGG, Metabolic pathway analysis, Genome annotation, Microalgae

1 Introduction

Chlorella sp. is one of the most popular microalgae applied in many industries like pharmaceuticals and health across many Asian countries for decades. These species have a unique ability to produce metabolites that possess various health-promoting activities such as antioxidant, antifungal, anticancer, and antiviral properties. They can also produce more pigments compared to other microalgal species and are used in various commercial applications due to their adaptability to different culture conditions.

Next-generation sequencing will continue to support *Chlorella* with improvement in metabolic flexibility via extensive gene and

Dhanasekaran Dharumadurai (ed.), *Chlorella Genomics, Cultivation, and Bioenterprise: Green Economy through Algal Innovations*, Springer Protocols Handbooks, https://doi.org/10.1007/978-1-0716-5276-3_16,
© The Author(s), under exclusive license to Springer Science+Business Media, LLC, part of Springer Nature 2026

enzyme networks contained in its genome. The complete genome sequence of many *Chlorella* spp. has been completed due to advancements in next-generation sequencing technology, which allows for a genome-wide assessment of the metabolic capabilities of *Chlorella*. Genome sequence data, when analysed exclusively, provides no biological information without functional annotation of the genes and identification of their integration into a metabolic network.

When functionally annotating genes using an orthology method, a researcher can determine the role of a gene by comparing it to a well-defined reference genome. A frequently used software tool is EggNOG-mapper, which associates genes with evolutionarily conserved orthologous groups and predicts their potential biochemical functions, including Gene Ontology, Enzyme Commission numbers, and KEGG Orthology IDs. Furthermore, the KEGG database is a curated source of the aforementioned functional annotations and provides a framework to map functional annotations to the various metabolic pathways and modules associated with a specific organism; thus, metabolic networks unique to each species may be reconstructed [1].

Using KEGG database resources to perform a metabolic pathway analysis of the genome sequences of *Chlorella* has led to the identification of the major metabolic pathways associated with the processes of carbon fixation, energy production, lipid biosynthesis, nitrogen assimilation, and secondary metabolite production in the organism [2]. This approach is extremely beneficial for determining the metabolic potential of *Chlorella*, establishing pathways that may limit productivity when attempting to manipulate metabolic engineering efforts. This chapter provides a comprehensive protocol for functional annotation and metabolic pathway reconstruction of the *Chlorella* genome using EggNOG-mapper and KEGG.

2 Materials

1. Predicted protein sequences of *Chlorella* genome (FASTA format).
2. EggNOG-mapper v2 web server.
3. KEGG Automatic Annotation Server (KAAS).
4. KEGG Mapper—Reconstruct Pathway tool.
5. Reference eukaryotic genomes in eggNOG and KEGG databases
6. Stable Internet connection and standard web browser.

3 Methodology

3.1 Functional Annotation Using EggNOG-Mapper

The EggNOG (Evolutionary Genealogy of Genes Ops) database is a valuable source of information for scientists who need to assign orthologous genes and functional categories to gene and protein sequences. EggNOG brings together the following types of information for individual gene/protein sequences: Orthologous Group Classifications and Annotating Functions, for example, using Gene Ontology (GO), KEGG, and the 2-syllable code (EC#). Additionally, this tool can help speed the process of annotating the Functions of a lot of different protein sequences by allowing researchers to easily find similar sequences already known to be in many of the four categories mentioned above (orthologous groups) [1].

3.2 EggNOG Web Interface Description

1. Access the EggNOG database homepage via https://eggnog.embl.de (Fig. 1).
2. The homepage displays an overview of the EggNOG v5.0 database, including the following:
   (a) Number of organisms covered.
   (b) Viral genomes included.
   (c) Orthologous groups.
   (d) Phylogenetic trees and alignments.

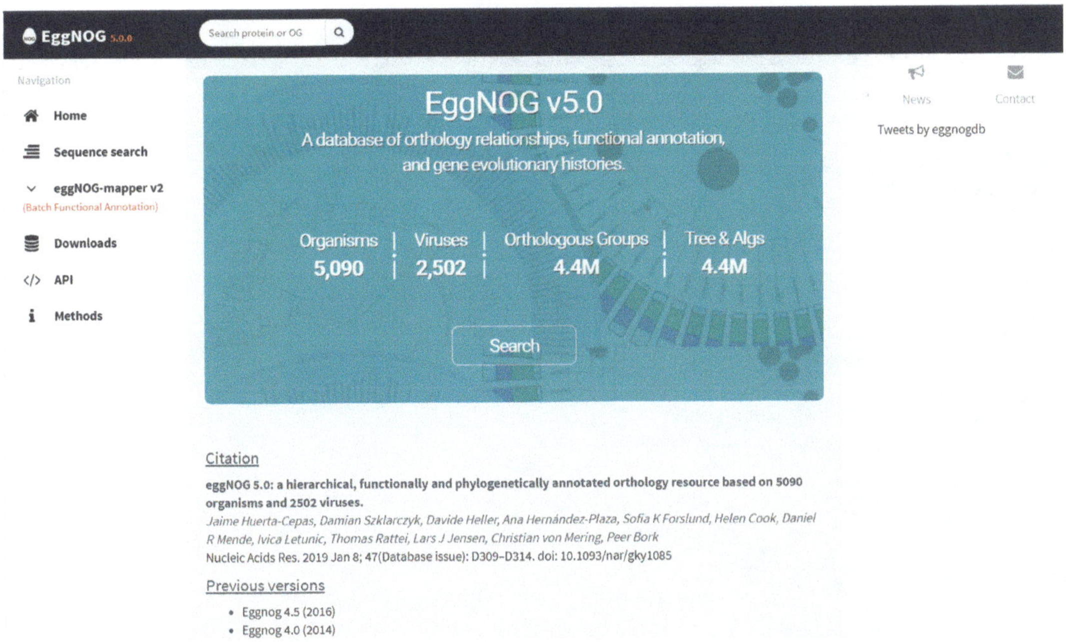

Fig. 1 Homepage of EggNOG v5.0 database

3. The left navigation panel provides access to the following:
   (a) Sequence search.
   (b) EggNOG-mapper v2 (Batch Functional Annotation).
   (c) Downloads.
   (d) API and methods.

3.3 Sequence Search (Single Protein Analysis)

1. Select "Sequence search" from the navigation panel to identify orthologs for a single protein sequence (Fig. 2).
2. Paste a single protein sequence in FASTA format or upload a FASTA file using the "Choose File" option.
3. Click Search to identify orthologous groups based on sequence similarity.
4. The output provides information on matched orthologs and functional annotations for individual proteins.

3.4 EggNOG-Mapper v2 (Batch Functional Annotation)

1. For genome-scale annotation, select EggNOG-mapper v2 under the "Batch Functional Annotation" section.
2. Upload predicted protein sequences of the *Chlorella* genome in FASTA format.
3. Select DIAMOND as the mapping algorithm for fast and accurate orthology assignment.
4. Choose the EggNOG v5.0 database and restrict taxonomy to Eukaryotes to improve annotation specificity.

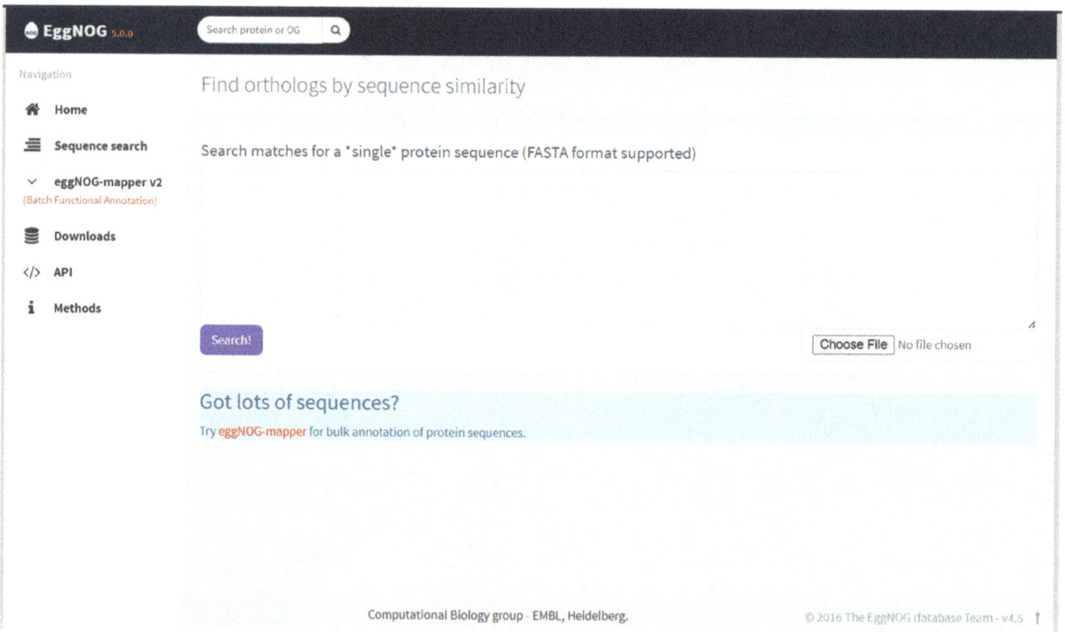

Fig. 2 EggNOG-mapper v2 interface for batch functional annotation

5. Enable functional annotation outputs including the following:
   (a) Gene Ontology (GO) terms.
   (b) KEGG Orthology (KO) identifiers.
   (c) Enzyme Commission (EC) numbers.
   (d) COG functional categories.
6. Retain default parameters for alignment coverage and e-value thresholds.
7. Submit the job and monitor the annotation progress via the results page.
8. Upon completion, download the annotation results in tab-delimited or Excel formats for downstream analyses.

3.5 KEGG-Based Functional Annotation and Metabolic Pathway Analysis

The functional annotation of the *Chlorella* genome and the reconstruction of its metabolic pathway were accomplished through KEGG annotation tools BlastKOALA, GhostKOALA, and KofamKOALA, which assign KO (KEGG Orthology) identifiers based on two criteria: sequence similarity and profile hidden Markov models (HMMs) [3] (Fig. 3).

3.6 BlastKOALA Analysis

1. Open the BlastKOALA web server available through the KEGG platform (https://www.kegg.jp/blastkoala/) [4] (Fig. 4).
2. Upload the predicted protein sequences of the *Chlorella* genome in FASTA format.
3. Select the appropriate taxonomy group (Eukaryotes) to improve annotation accuracy.
4. Choose the default BLAST-based search parameters provided by the server.
5. Submit the job and monitor the progress using the job identifier provided by the server.
6. Upon completion, download the output file containing assigned KEGG Orthology (KO) identifiers, functional descriptions, and pathway associations.

3.7 GhostKOALA Analysis

1. Access the GhostKOALA web server (https://www.kegg.jp/ghostkoala/) through the KEGG annotation tools section [4] (Fig. 5).
2. Upload the *Chlorella* predicted protein sequences in FASTA format.
3. Select GHOSTX as the homology search algorithm, which is optimized for large-scale and rapid sequence comparisons.
4. Choose Eukaryotes as the taxonomic category to restrict the reference dataset.

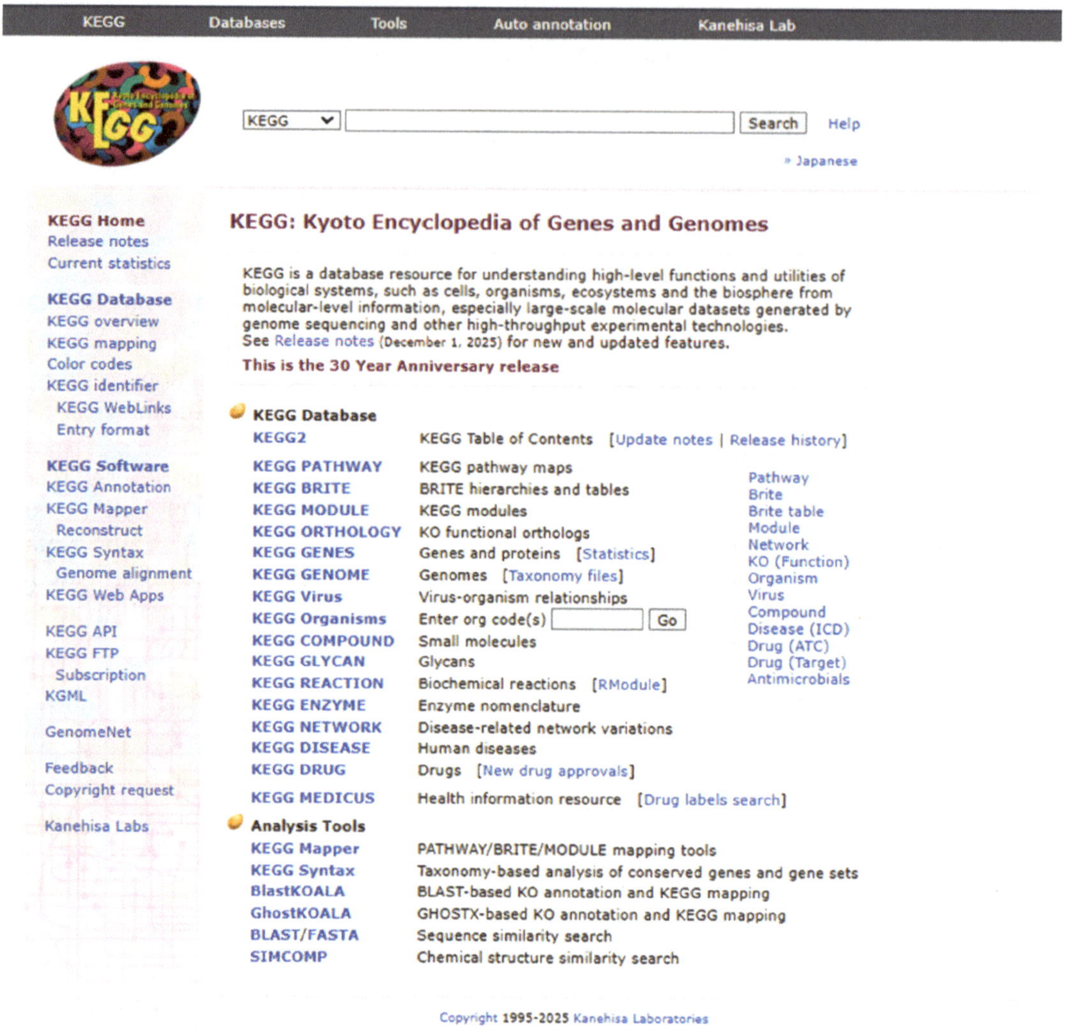

Fig. 3 KEGG web interface displaying databases and annotation tools for functional and metabolic pathway analysis

5. Submit the job and retrieve the KO assignment results after completion (Fig. 7).
6. Download the annotation output containing KO numbers and functional classifications.

3.8 KofamKOALA Analysis

1. Open the KofamKOALA web server (https://www.genome.jp/tools/kofamkoala/) [5] (Fig. 6).
2. Upload the *Chlorella* protein sequences in FASTA format.
3. Enable HMM-based KO assignment using predefined adaptive score thresholds.

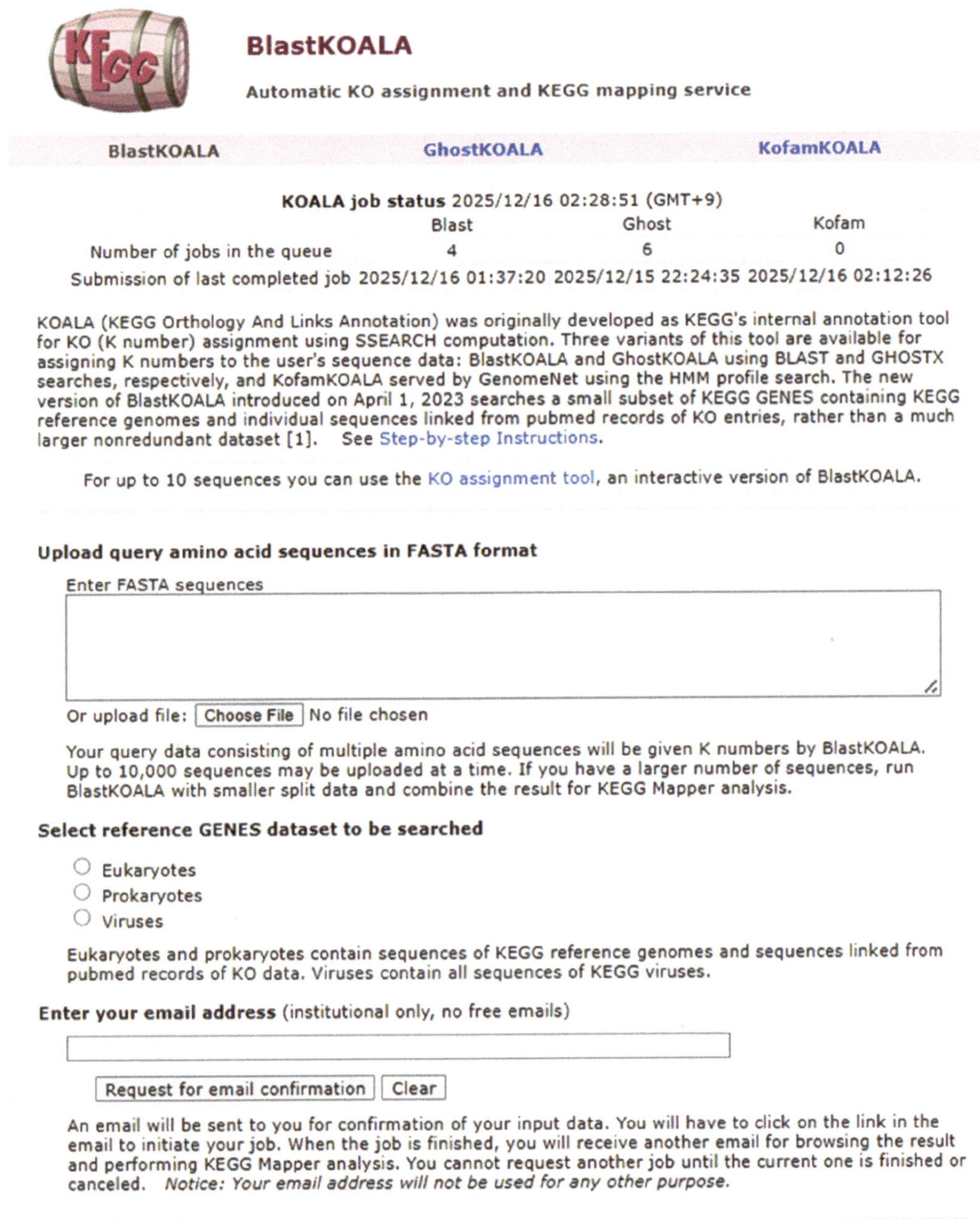

Fig. 4 BlastKOALA interface for KEGG Orthology assignment

4. Retain default threshold settings to minimize false-positive annotations.
5. Submit the analysis and download the results containing KO identifiers with confidence scores.

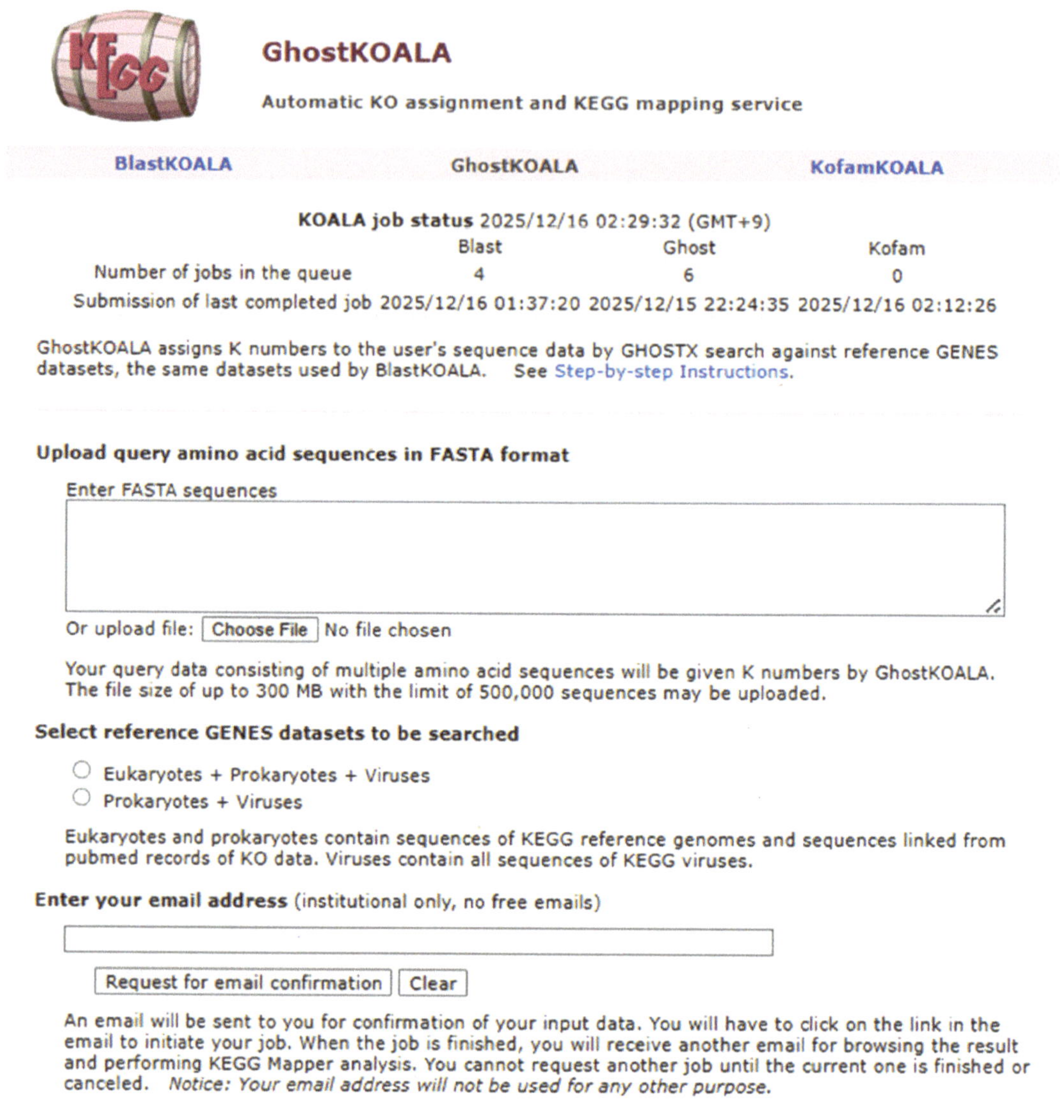

Fig. 5 GhostKOALA interface using GHOSTX algorithm

3.9 Data Interpretation and Visualization

1. The completeness of metabolic pathways and KEGG modules is evaluated based on the presence of essential KO components.
2. A pathway map generated by KEGG Mapper is used to identify key enzymes and metabolic bottlenecks.
3. A functional annotation from EggNOG-mapper and KEGG is cross-validated to enhance annotation confidence.

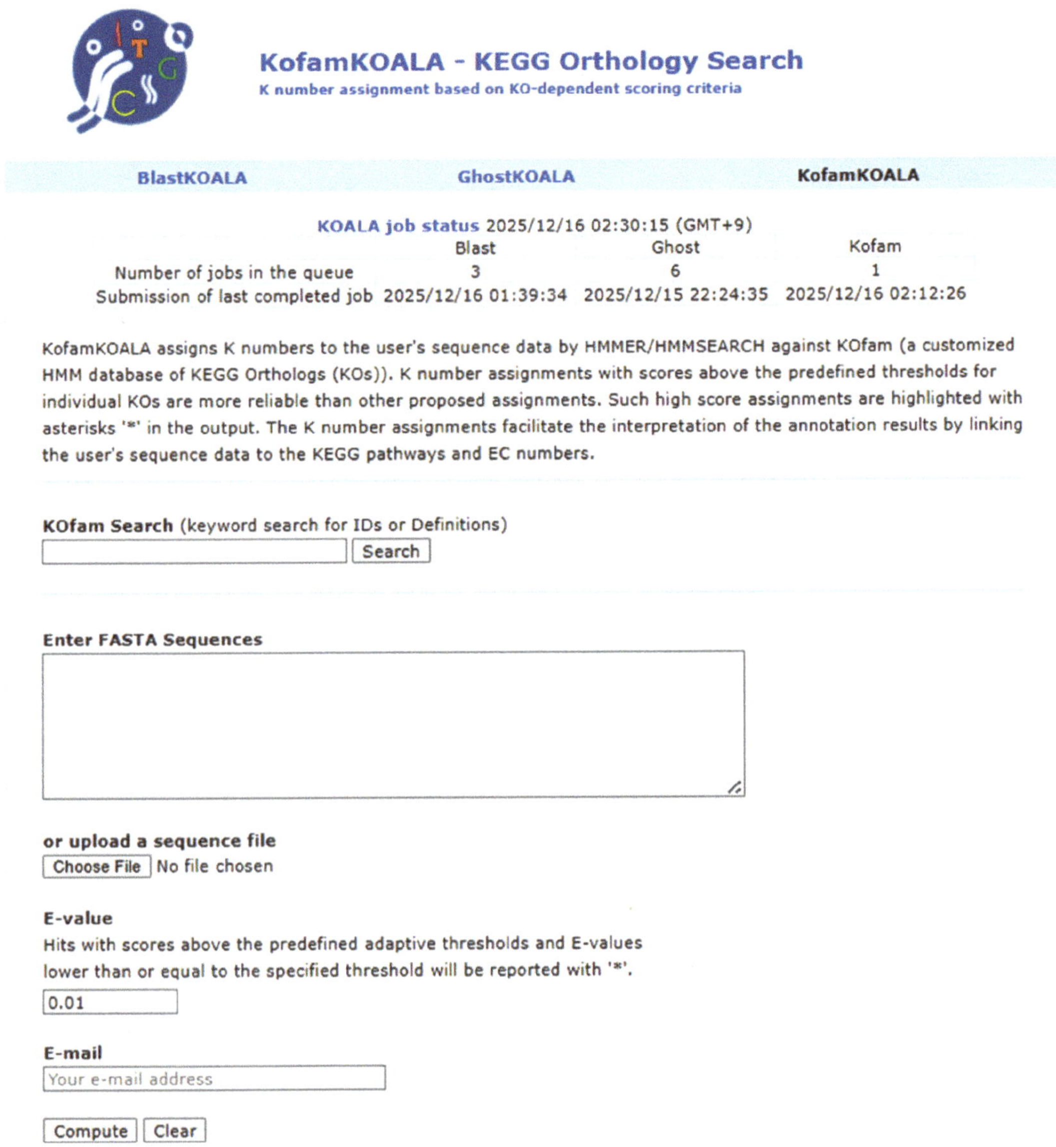

Fig. 6 KofamKOALA interface for HMM-based KO assignment

4 Notes

1. Use high-quality, nonredundant protein sequences for accurate annotation.
2. Ensure a stable internet connection throughout the analysis.
3. KEGG and EggNOG annotations are predictive and require experimental validation.
4. Employing multiple KO assignment tools improves annotation reliability.

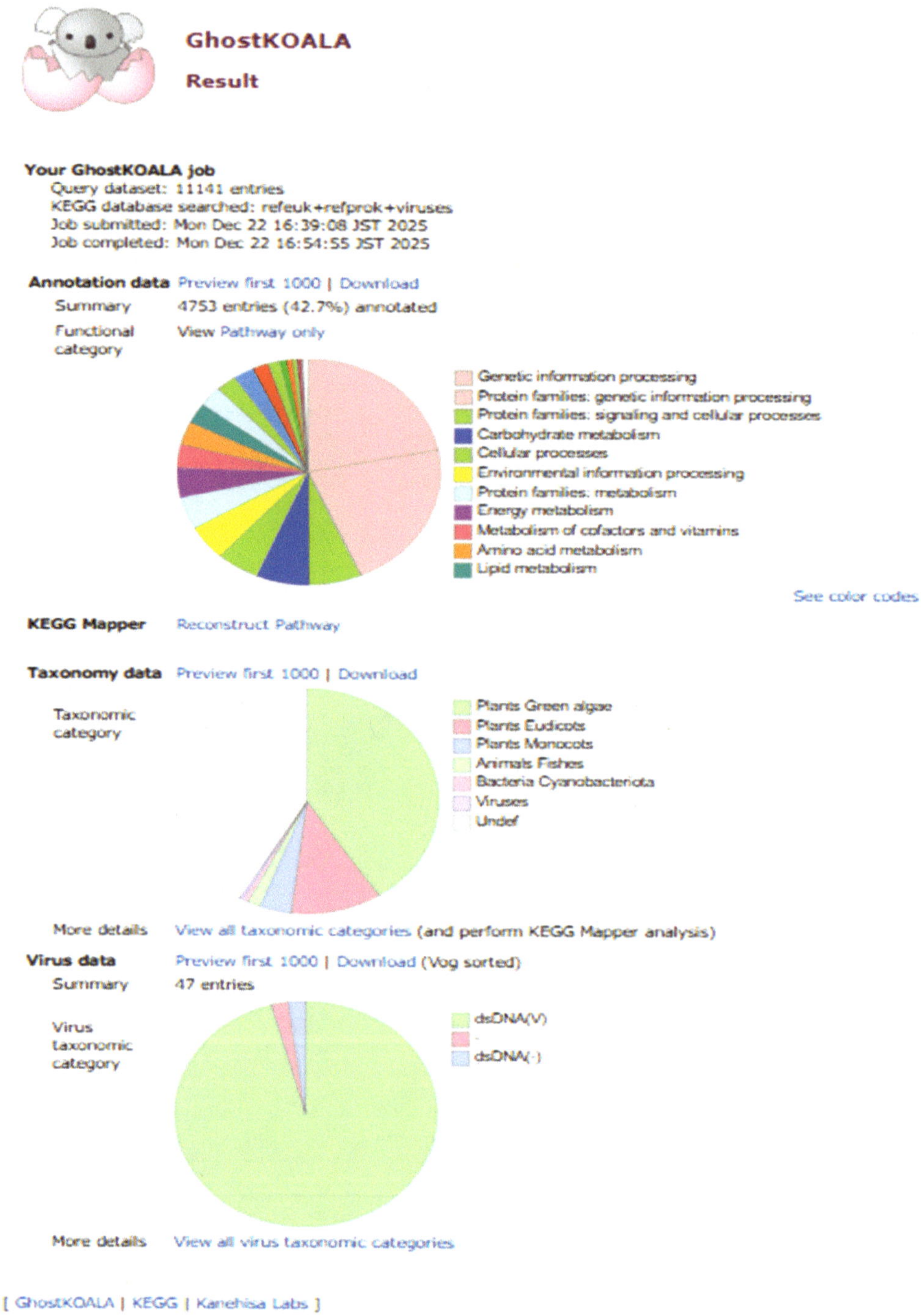

Fig. 7 Overview of GhostKOALA Functional and Taxonomic Annotation Results

Acknowledgments

We sincerely acknowledge the financial support from the DBT (Govt. of India), for the establishment of the National Repository for Microalgae and Cyanobacteria—Freshwater & Marine (NRMC-F & M) (Phase II) (BT/PR29901/PBD26/694/2018). We gratefully acknowledge the Chief Minister Research Grant (CMRG), Govt. of Tamil Nadu for providing financial support (CMRG 2023-2024 Project No. CMRG2023BMBO7047), Rashtriya Uchchatar Shiksha Abhiyan (RUSA) – 2.0 Biological Sciences (TN RUSA: 311/RUSA (2.0)/2018 Dt. December 02, 2020) and School of Life Sciences, Bharathidasan University, Tiruchirappalli, Tamil Nadu, India.

References

1. Huerta-Cepas J, Szklarczyk D, Heller D, Hernández-Plaza A, Forslund SK, Cook H, Mende DR, Letunic I, Rattei T, Jensen LJ, von Mering C, Bork P (2019) eggNOG 5.0: a hierarchical, functionally and phylogenetically annotated orthology resource based on 5090 organisms and 2502 viruses. Nucleic Acids Res 47(D1):D309–D314. https://doi.org/10.1093/nar/gky1085
2. Kanehisa M, Furumichi M, Sato Y, Ishiguro-Watanabe M, Tanabe M (2021) KEGG: integrating viruses and cellular organisms. Nucleic Acids Res 49(D1):D545–D551
3. Zhang C, Chen Z, Zhang M, Jia S (2023) KEGG_extractor: an effective extraction tool for KEGG orthologs. Genes 14(2):386. https://doi.org/10.3390/genes14020386
4. Kanehisa M, Sato Y, Morishima K (2016) BlastKOALA and GhostKOALA: KEGG tools for functional characterization of genome and metagenome sequences. J Mol Biol 428(4):726–731. https://doi.org/10.1016/j.jmb.2015.11.006
5. Aramaki T, Blanc-Mathieu R, Endo H, Ohkubo K, Kanehisa M, Goto S, Ogata H (2020) KofamKOALA: KEGG Ortholog assignment based on profile HMM and adaptive score threshold. Bioinformatics 36(7):2251–2252. https://doi.org/10.1093/bioinformatics/btz859

Chapter 17

CRISPR-Cas9 Guide RNA Designing for Genome Editing in *Chlorella* sp. using CHOPCHOP

Anandhi Anandaraj and Rajasekar Thirunavukkarasu

Abstract

Clustered Regularly Interspaced Short Palindromic Repeat (CRISPR) and CRISPR-associated (CRISPR-Cas) genes system is an emerging genome editing tool that play an important role in the diverse array of the organisms and cell types. The methods were primarily developed to induce targeted mutations in DNA. A suitable Cas protein and guide RNA are necessary for CRISPR editing to be effective. CHOPCHOP is a web tool for identifying the CRISPR-Cas Single guide RNA (sgRNA) target. The Major use of CHOPCHOP extend our toolbox beyond Knockouts. Here we summarize the methods focusing primarily on the computational tools used to design the single guide RNA for *Chlorella* sp. for CRISPR editing.

Key words CRISPR-Cas, Genome editing , CHOPCHOP, Single guide RNA, *Chlorella* sp., Modern biology

1 Introduction

CRISPR-Cas is a recent molecular technique for modern biology used for introducing repair-induced mutations genes for activation or repress and transcription as an enrichment tool for the sequencing [1, 2] of targeted mutation. The CRISPR/Cas system was first adapted for genome editing in 2012 [3]. It is used as a diagnostic tool to complete the whole organism's lineage tracing to target RNA molecules for destruction or editing to tract transcript in live cells [4].

CRISPR-Cas is the most commonly used webtool for well-organized genome editing of bacteria, plant, cell lines, and tissues [5]. Different experimental setup and methods can be used to deliver CRISPR system. Genome editing tool in the cultured cells, either transfection or electroporation, is frequently used to deliver a plasmid covering a ubiquitous promoter driving Cas9 and SgRNA expression [6].

Dhanasekaran Dharumadurai (ed.), *Chlorella Genomics, Cultivation, and Bioenterprise: Green Economy through Algal Innovations*, Springer Protocols Handbooks, https://doi.org/10.1007/978-1-0716-5276-3_17,
© The Author(s), under exclusive license to Springer Science+Business Media, LLC, part of Springer Nature 2026

Development of genetic manipulation tools for *Chlorella* sp. remains very limited [7]. Efforts have been made to establish gene over expression and CSISPR-mediated gene editing in *Chlorella* sp. [8]. CRISPR/Cas9-mediated genome editing using plant-derived vectors in two freshwater *Chlorella* strain showed these methods have been unstable with gene knockout efficiencies too low to be used in practice. The genetic modification methods are unstable and nonreplicative across the different *Chlorella* strains, and transgenes are typically missing after a few generations. Therefore, it is very important to develop stable and high-efficiency CRISPR-Cas9-based genome editing tools for *Chlorella*.

Mechanism of gene editing in *Chlorella* sp., the CRISPR array is replaced by a synthetic construct which constitutes a single guide RNA (sgRNA) and a Cas enzyme. The small single guide RNA is an RNA molecule that is usually a custom-designed sequence containing approximately 20 nucleotides which is complementary to the target site of the eukaryotic genome and the Cas enzyme is the scissors that cuts the target DNA when the sgRNA forms a complex upon binding with the Cas system. A short sequence PAM (Protospacer Adjacent Motif) is required to bind the sgRNA with the Cas enzyme where it locks onto the DNA and creates a double-stranded break at the precise location specified by the guide RNA. Once the cut is made, the eukaryote cells activate the DNA repair machinery, which helps in achieving gene editing either by Nonhomologous end joining pathway for knockout (inactivate a gene) or by Homology Directed Repair pathway to knock in (insert a gene) [9].

When using CRISPR/Cas9 for genome editing, one needs to express a gRNA designed to direct Cas9 to their target sequence of choice and their preferred Cas9 variant (with appropriate PAM sequence) to modify the genomic locus. Target selection mainly based on presence of PAM (protospacer adjacent motif) which provides a broad range of potential target sites and gRNA. Randomly selecting the gRNA can reduce efficiency, therefore optimizing gRNA design is essential to enhance target specificity and editing efficiency.

2 Materials

1. *Chlorella* sp. Genome
2. Online tool CHOPCHOP

3 Methods

For the genome editing in *Chlorella* sp. using CHOPCHOP, the following steps are followed. Online tool CHOPCHOP (https://chopchop.cbu.uib.no) is used here to design the single guide RNA.

3.1 Input

Step 1: Define your target

Enter the specific gene name or accession number in the "Target" box (if the organism genome annotation is already loaded for the organism) or paste the DNA sequence (FASTA format) when gene annotations are not found (e.g., *Chlorella sorokiniana* CS12_1230) Fig. 1.

Step 2: Choose CRISPR System and set the goal

Choose the enzyme which needs to be applied and the type of editing need to be performed (Fig. 2).

Step 3: Run "Find Target Sites"

After choosing the type of editing that needs to be carried out, click "*Find target sites.*"

In the "*options*" menu, if needed requirements can be adjusted as per the need (Figs. 3 and 4).

3.2 Output

Step 4: Based on the specified parameters, the target sequences are displayed as per the rank order mentioning the target sites' genomic location, strand, GC content, self-complementary, and efficiency. Further details of the target sequence like primers, its

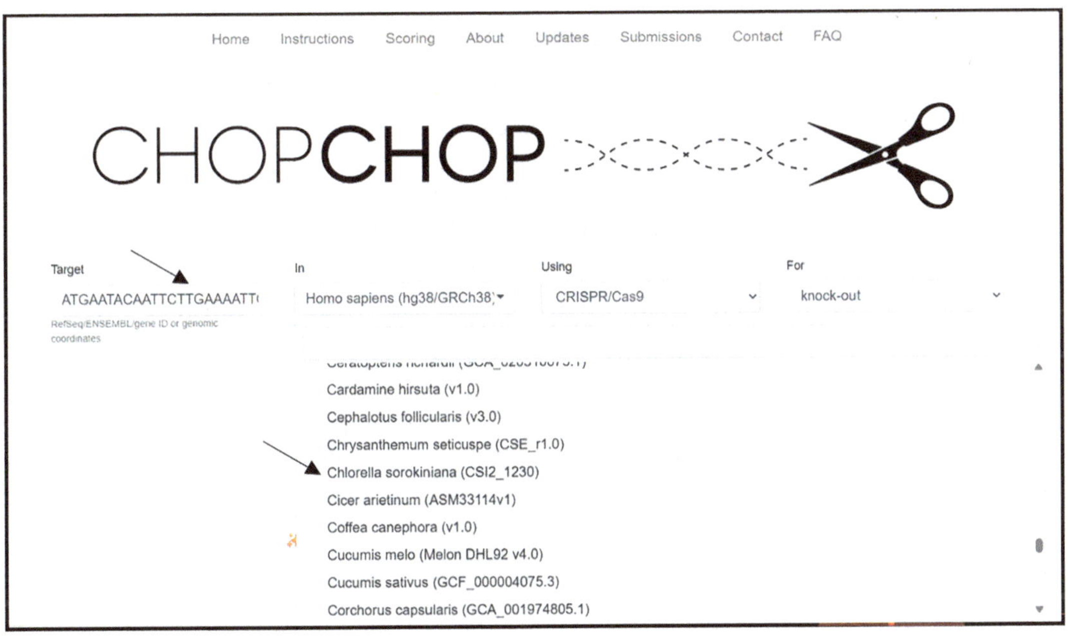

Fig. 1 Enter the specific gene name in the CHOPCHOP tool

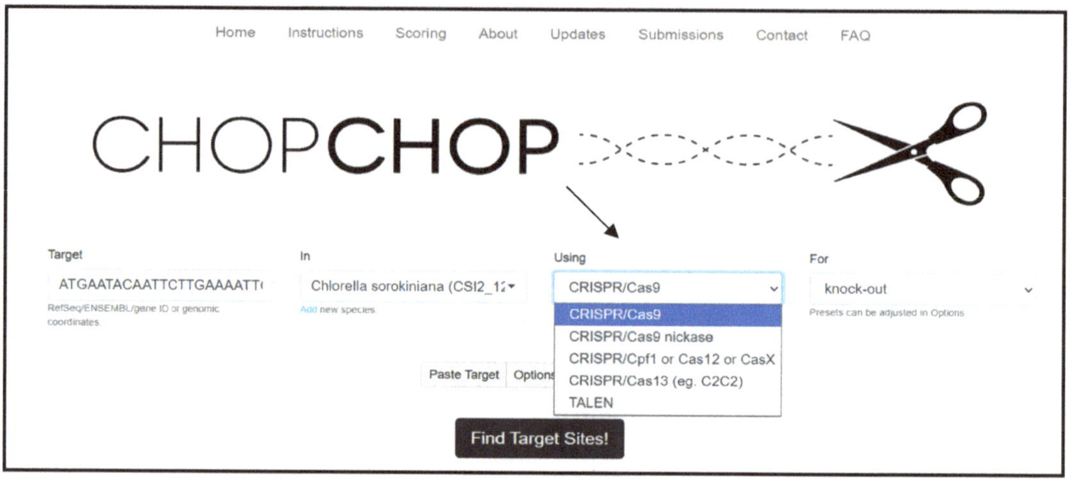

Fig. 2 Find the target sites in the genome using CHOPCHOP tool

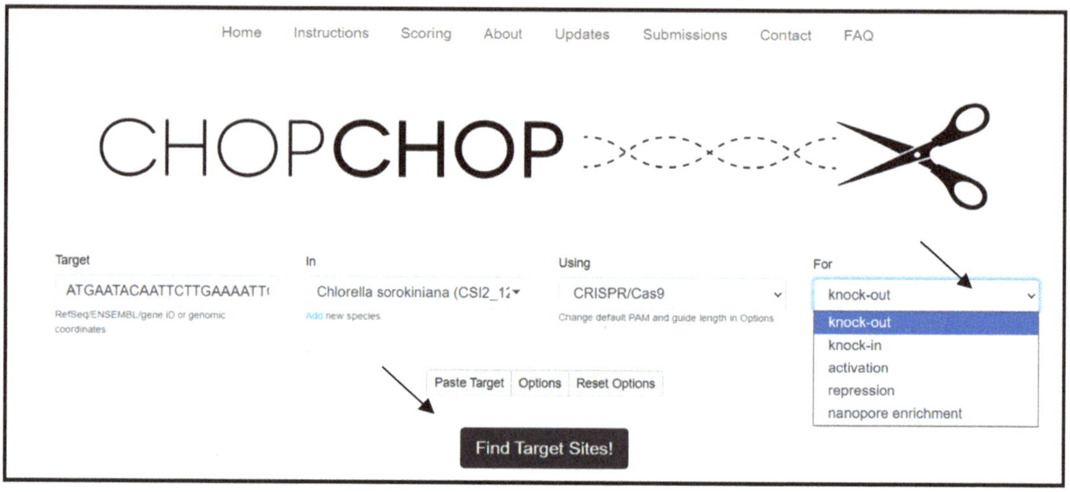

Fig. 3 Run the target sites in the genome using CHOPCHOP tool

melting temperatures, off targets, and product size can be obtained by clicking the sequence. The portion of sequence in red color denotes the PAM sites (Fig. 5).

Step 5: Validation of the result using IDT Design Checker (optional)

Optimizing gRNA design is essential to enhance target specificity and maximize editing efficiency. It is suggested to take the top-ranked sequences from CHOPCHOP and run them through a secondary tool like IDT Design checker to cross verify the on-target and off target scores before ordering the guides (Fig. 6).

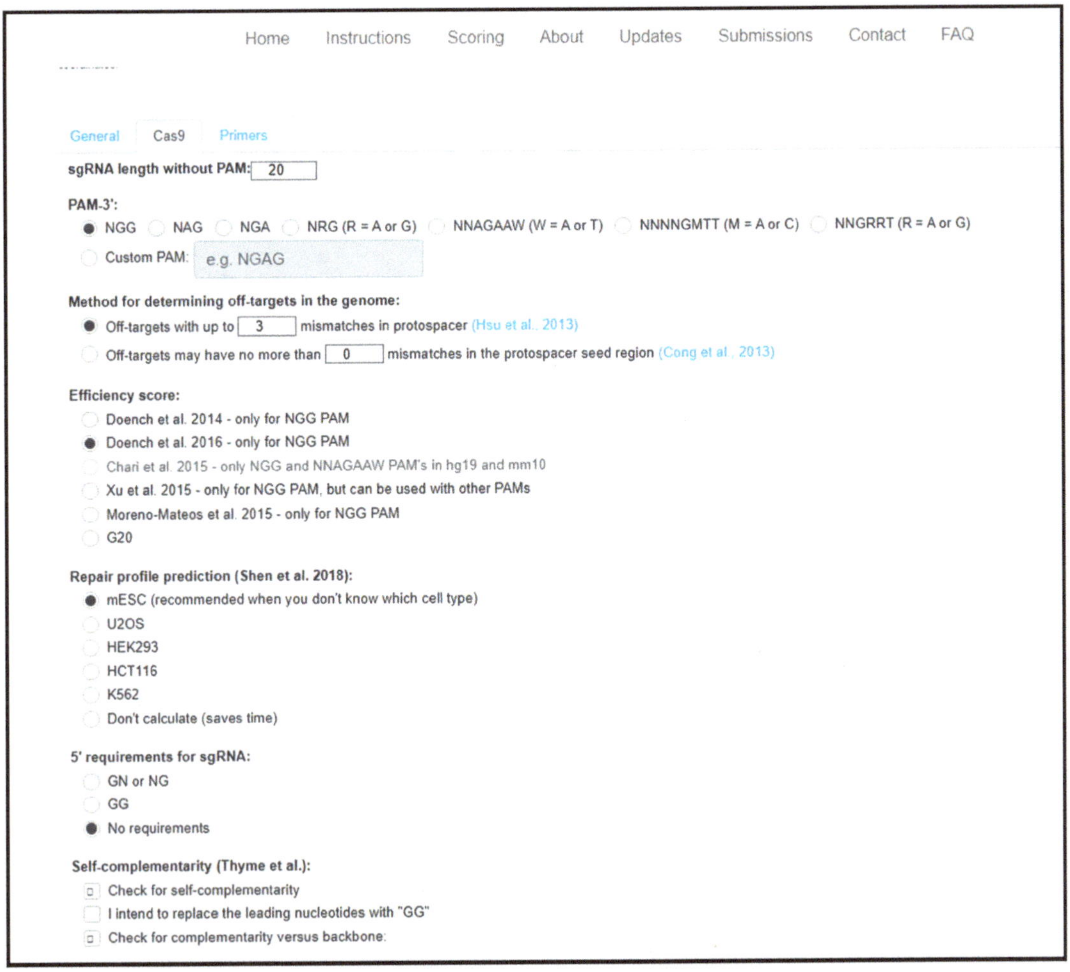

Fig. 4 The output of target sites in the genome using CHOPCHOP tool

4 Notes

CHOPCHOP displays several sequences based on parameters provided in input. To pick the best, the following criteria can be considered:

1. Rank: Usually, the first few results are high rated.
2. GC content: Ideally choose GC content between 40% and 70%.
3. Off target Score: A relevant guide should have zero mismatches, which is the most critical factor.
4. Self-Complementarity: A high score denotes that guide might fold onto itself.
5. Efficiency Score: Prefer for high predicted efficiency score.

Fig. 5 The target sites' genomic location, strand output data from genome using CHOPCHOP tool

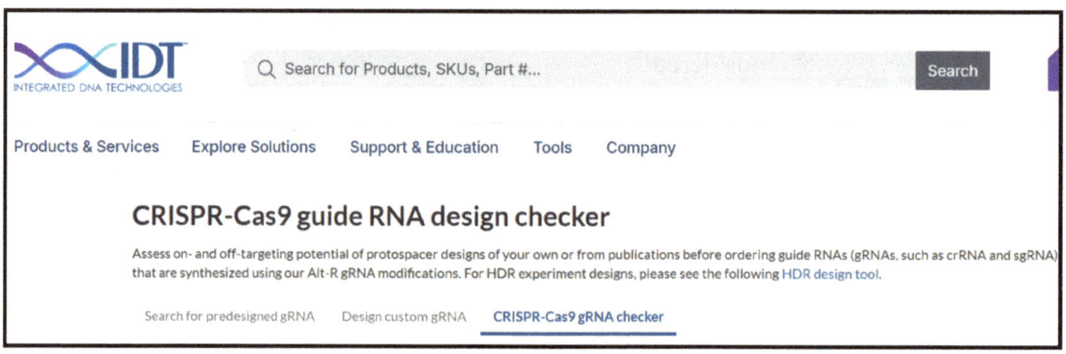

Fig. 6 Optimizing gRNA design checker

References

1. Labun K, Montague TG, Krause M, Torres Cleuren YN, Tjeldnes H, Valen E (2019) CHOPCHOP v3: expanding the CRISPR web toolbox beyond genome editing. Nucleic Acids Res 47(W1):W171–W174
2. Gabrieli T, Sharim H, Fridman D, Arbib N, Michaeli Y, Ebenstein Y (2018) Selective nanopore sequencing of human BRCA1 by Cas9-assisted targeting of chromosome segments (CATCH). Nucleic Acids Res 46(14):e87–e87
3. Jinek M, Chylinski K, Fonfara I, Hauer M, Doudna JA, Charpentier E (2012) A programmable dual-RNA–guided DNA endonuclease in adaptive bacterial immunity. Science 337(6096): 816–821
4. Abudayyeh OO, Gootenberg JS, Essletzbichler P, Han S, Joung J, Belanto JJ et al (2017) RNA targeting with CRISPR–Cas13. Nature 550(7675):280–284
5. Niu Y, Shen B, Cui Y, Chen Y, Wang J, Wang L et al (2014) Generation of gene-modified cynomolgus monkey via Cas9/RNA-mediated gene targeting in one-cell embryos. Cell 156(4):836–843
6. Cong L, Ran FA, Cox D, Lin S, Barretto R, Habib N et al (2013) Multiplex genome engineering using CRISPR/Cas systems. Science 339(6121):819–823
7. Ajjawi I, Verruto J, Aqui M, Soriaga LB, Coppersmith J, Kwok K et al (2017) Lipid production in Nannochloropsis gaditana is doubled by decreasing expression of a single transcriptional regulator. Nat Biotechnol 35(7):647–652
8. Kim J, Chang KS, Lee S, Jin E (2021) Establishment of a genome editing tool using CRISPR-Cas9 in Chlorella vulgaris UTEX395. Int J Mol Sci 22(2):480
9. Ng IS, Keskin BB, Tan SI (2020) A critical review of genome editing and synthetic biology applications in metabolic engineering of microalgae and cyanobacteria. Biotechnol J 15(8): 1900228

Part IV

Mass Cultivation of *Chlorella* spp.

Chapter 18

Comparative Genome Analysis of *Chlorella* Species

Asraf Sithikka Rasheed, Akilan Elumalai, Thajuddin Nooruddin, and Dhanasekaran Dharumadurai

Abstract

Genome analysis of microalgae is crucial for the comprehension of species diversity, interpretation of their metabolic pathways, which facilitates genetic engineering. The comparative analysis of the *Chlorella* genome provides significant information on their functional diversity, evolutionary relationships, and adaptability for a broad spectrum of environmental and biotechnological applications. The NCBI genome assembly database offers a wide range of genomic interference thereby providing its accession IDs with substantial metadata such as various statistics and wide range of assemblies from simple scaffolds to complex structures, thereby serving as significant source for the comparative genome analysis.

Key words NCBI, *Chlorella* genome, Genomic interference, Comparative genome analysis

1 Introduction

In the rapidly advancing domain of genomics, genome assembly, annotation, and comparative genomics have emerged as crucial approaches which enables researchers to interpret the code of genetics [1]. Comparative genomics is an evolving branch of genomics that analyses the genetic data within and across the species, resulting in novel perspectives into various biological studies. Yet, a wide range of technological and data-related issues prevent researchers from fully exploiting its potential [2]. This concept leverages the computational methods and evolutionary theory to interpret the genomic sequences. This approach allows researchers to identify the functional elements, such as dynamic regions or regulatory sites, while developing hypotheses that may be verified through experimental studies. Thus, comparative genomics provides an evolving opportunity to explore the evolutionary history of organisms and their distinctive characteristics as whole genome sequencing becomes increasingly accessible [3]. The genome assembly database of National Centre for Biotechnology

Dhanasekaran Dharumadurai (ed.), *Chlorella Genomics, Cultivation, and Bioenterprise: Green Economy through Algal Innovations*, Springer Protocols Handbooks, https://doi.org/10.1007/978-1-0716-5276-3_18,
© The Author(s), under exclusive license to Springer Science+Business Media, LLC, part of Springer Nature 2026

Information (NCBI) offers permanent accession IDs and enables comprehensive data monitoring of the genome assemblies. The database design is highly versatile and flexible to deal with a broad spectrum of genome assemblies encompassing simple sets of disorganized sequences of scaffolds or contigs, whole single chromosome of bacterial genome, or a complicated genome structure of a complex organism like a human with intricate allelic variants. The assembly database serves as a source for comprehensive assembly metadata, including assembly ID, basic statistical data of the assembly such as total length and gap of the sequence, number of scaffolds and contigs, as well as the contiguity measures like N50 along with the updated assemblies. Besides, this database monitors the relationship between the assemblies submitted to the International Nucleotide Sequence Database Consortium (INSDC) and the assemblies published in the NCBI RefSeq project [4]. This chapter provides a simple structural protocol for the comparative genome analysis of *Chlorella*.

2 Materials

1. Hardware resources with sufficient storage.
2. MicroSoft Office package (MS Word and MS Excel).
3. Stable Internet.
4. *Chlorella* genome.

3 Methodology

1. Go to the genome assembly database (https://www.ncbi.nlm.nih.gov/datasets/genome/) in NCBI.
2. Search for the desired genus or species name in the search box (6 *Chlorella* strains are given as examples).
3. The result table includes encompasses the assembly ID, Genbank ID, RefSeq, Species name, strain or isolate name, along with the sidebar for assembly details and download options.
4. Click the assembly ID, which provides all the essential information of the genome assembly including their publishing data, assembly statistics, assembly methods, sample details, and chromosome details.
5. Collect these data from various species or strain specific genomes, organize the acquired data in an excel sheet (Fig. 1), and compare their genomic interference through their genome sizes (Fig. 2), No. of Scaffolds in each genome (Fig. 3), and through the GC % of the genome (Fig. 4) to evaluate their evolutionary relationships and much more (*see* **Notes 1–4**).

	A	B	C	D	E	F	G	H	I
1	Species name	Strain name	Genbank ID	Genome size	No. of Scaffolds	No. of contigs	N50	L50	GC %
2	*Chlorella sorokiniana*	SLA-04	GCA_025917655.1	3,86,12,678	13	13	3 Mb	5	66
3	*Chlorella vulgaris*	211/11P	GCA_023343905.1	4,01,80,792	43	55	1.8 Mb	8	62
4	*Chlorella variabilis*	NC64A	GCA_000147415.1	4,61,59,512	414	3,810	27.9 kb	432	67
5	*Chlorella ohadii*	1	GCA_025026875.1	5,70,91,209	486	3,304	31.7 kb	531	64
6	*Chlorella sorokiniana*	1230	GCA_003130725.1	5,83,62,567	18	20	3.8 Mb	7	64
7	*Chlorella vulgaris*	PKVL7422	GCA_037855405.1	4,06,58,414	91	91	1 Mb	14	61.5

Fig. 1 Collection of genomic data from NCBI platform

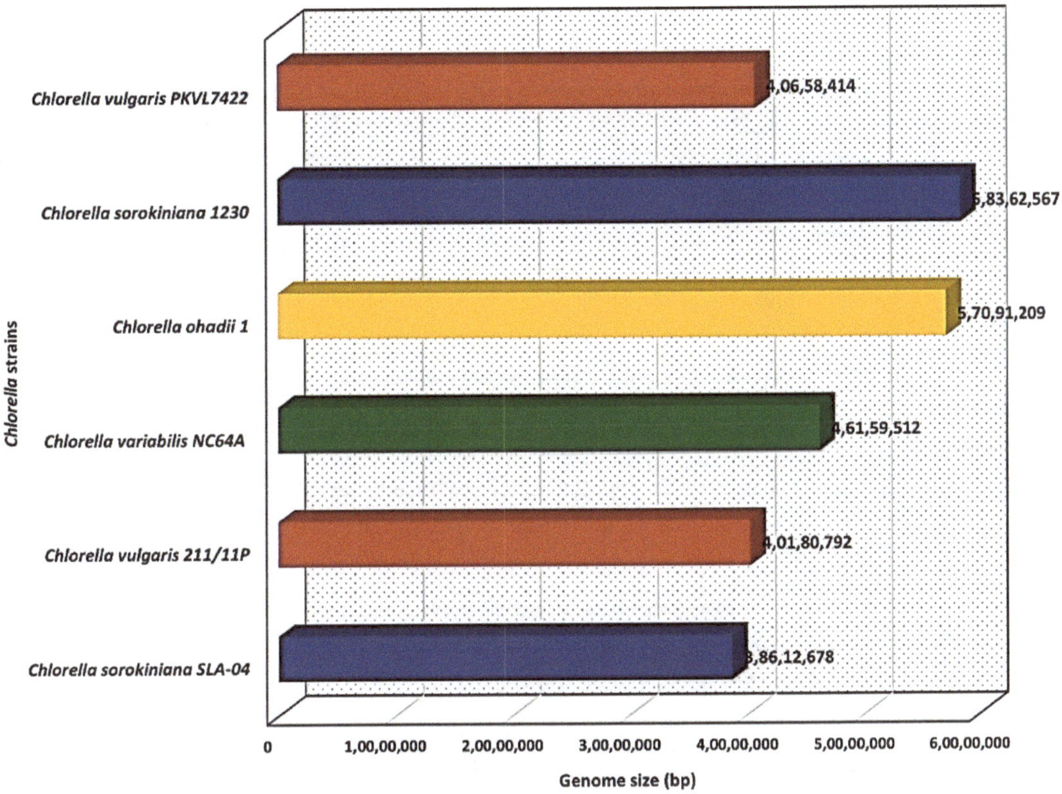

Fig. 2 Comparative analysis of genome through the genome sizes

4 Notes

1. Maintain a clear and structured folders to store the downloaded genomes and related files, separately.
2. Employ correct and consistence naming system for all the files (for the genomes—species_strain_accession).
3. Compare the genomes based on same set of the parameters for reliability.
4. Cross check the data through both the NCBI GenBank and RefSeq for verification.

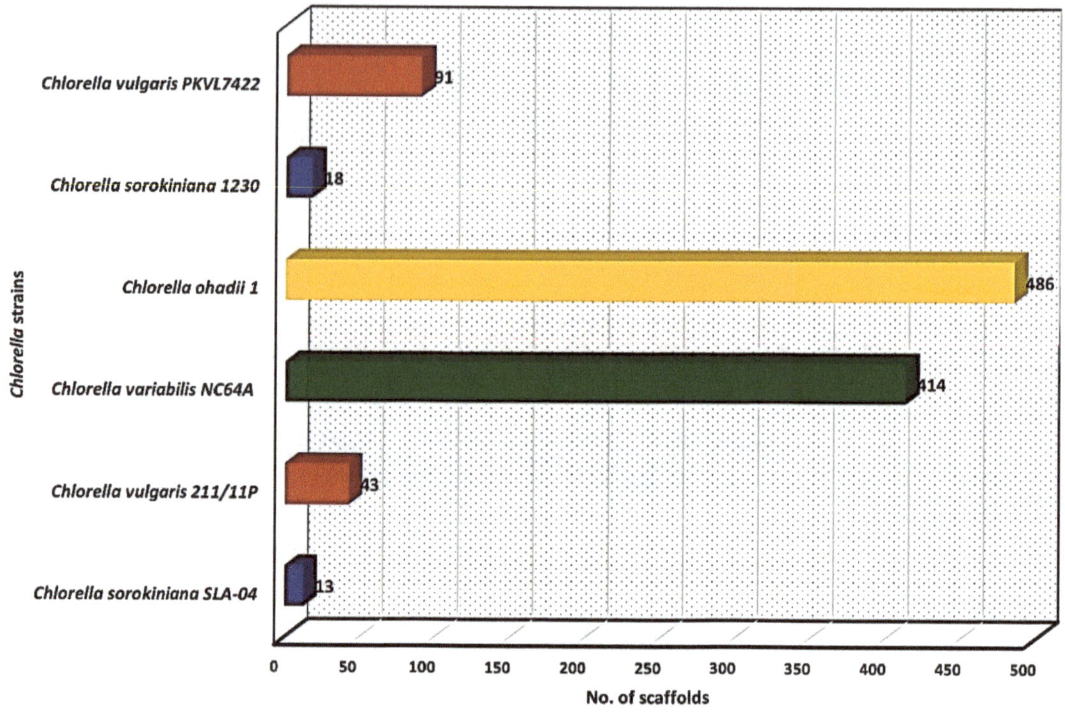

Fig. 3 Comparative analysis of genome through the No. of Scaffolds in each genome

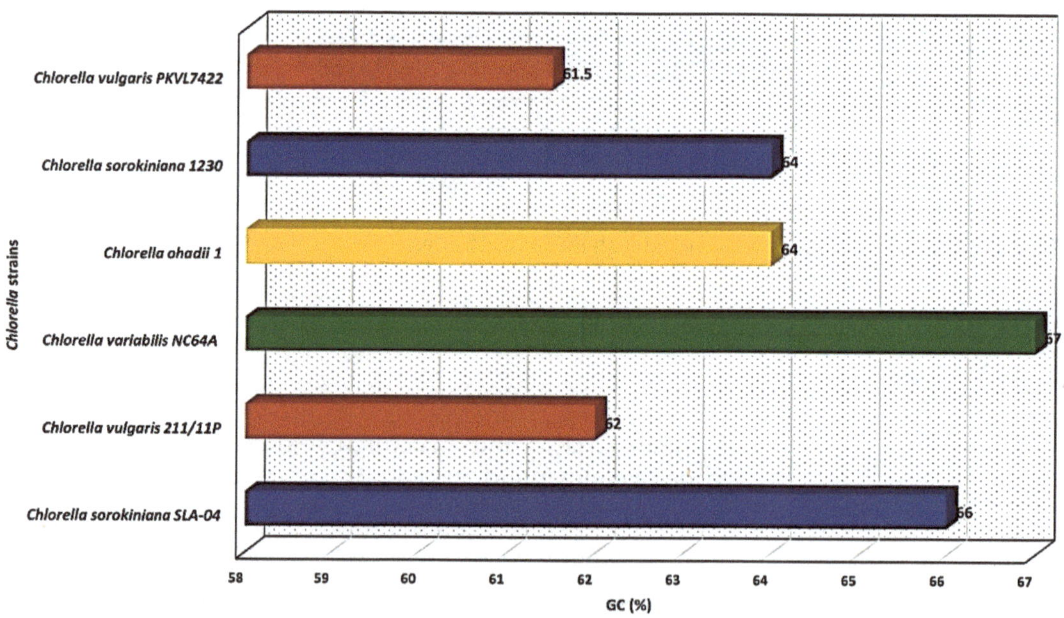

Fig. 4 Comparative analysis of genome through the GC % of each genome

Acknowledgments

We sincerely acknowledge the financial support from the DBT (Govt. of India), for the establishment of National Repository for Microalgae and Cyanobacteria—Freshwater & Marine (NRMC-F & M) (Phase II) (BT/PR29901/PBD26/694/2018). We gratefully acknowledge the Chief Minister Research Grant (CMRG), Govt. of Tamil Nadu for providing financial support (CMRG 2023-2024 Project No. CMRG2023BMBO7047), Rashtriya Uchchatar Shiksha Abhiyan (RUSA) – 2.0 Biological Sciences (TN RUSA: 311/RUSA (2.0)/2018 Dt. December 02, 2020) and School of Life Sciences, Bharathidasan University, Tiruchirappalli, Tamil Nadu, India.

References

1. Tang H, Krishnakumar V, Zeng X, Xu Z, Taranto A, Lomas JS, Zhang Y, Huang Y, Wang Y, Yim WC, Zhang J (2024) JCVI: a versatile toolkit for comparative genomics analysis. iMeta 3(4):e211
2. Bornstein K, Gryan G, Chang ES, Marchler-Bauer A, Schneider VA (2023) The NIH comparative genomics resource: addressing the promises and challenges of comparative genomics on human health. BMC Genomics 24(1):575
3. Stajich, J.E., 2013. Comparative genomics (pp. 380–386). Princeton: Princeton University Press
4. Kitts PA, Church DM, Thibaud-Nissen F, Choi J, Hem V, Sapojnikov V, Smith RG, Tatusova T, Xiang C, Zherikov A, DiCuccio M (2016) Assembly: a resource for assembled genomes at NCBI. Nucleic Acids Res 44(D1):D73–D80

Chapter 19

Preparation of Starter Culture of *Chlorella*

Priyanka Jayam Rajendran, Fayaazuddin Thajuddin, and Dhanasekaran Dharumadurai

Abstract

In this protocol, BG-11 is used to generate and scale up a *Chlorella* sp. starter culture under regulated laboratory circumstances. The goal is to cultivate a healthy, contamination-free microalgal culture that may be utilized to produce biomass and conduct additional studies. To promote rapid growth, the culture is maintained under optimal light, temperature, and aeration conditions. The growth is measured by optical density and using a standard conversion factor to determine biomass concentration. The starting culture has consistent growth patterns, making it ideal for large-scale cultivation and biochemical analysis.

Key words BG-11 medium, Optical density, *Chlorella* sp., Starter culture

1 Introduction

Chlorella are eukaryotic photosynthetic microorganisms that use sunlight, carbon dioxide, and water to produce biomass. This biomass can be used to create useful products (such as biofuels and food supplements). Microalgae can spread rapidly due to their ability to flourish in a wide range of conditions and their basic structure. This photosynthetic organism is a preferable source compared to other plant sources because microalgae can grow in varying amounts of nutrients and can adapt to changes in light and temperature. The culture medium is a nutrient-rich solution that promotes the growth of microorganisms and cells. Nitrogen, phosphorus, trace metals, and vitamins are among the common components of the medium. The composition of the culture influences the quality and quantity of fatty acids synthesized, as well as the quantities and properties of other compounds [2]. The development of an early cultivation is an essential step to obtaining a healthy and homogenous *Chlorella* biomass for subsequent investigations. A well-established starter culture provides continuous growth, contamination-free inoculum, and repeatable outcomes in large-scale

Dhanasekaran Dharumadurai (ed.), *Chlorella Genomics, Cultivation, and Bioenterprise: Green Economy through Algal Innovations*, Springer Protocols Handbooks, https://doi.org/10.1007/978-1-0716-5276-3_19,
© The Author(s), under exclusive license to Springer Science+Business Media, LLC, part of Springer Nature 2026

cultivation. This chapter discusses the supplies, procedures, and parameters used to create a pure and rapidly growing *Chlorella* starter culture.

2 Materials

1. *Chlorella* sp. pure isolate.
2. BG-11 medium (liquid and agar).
3. Erlenmeyer flasks (250 mL, 500 mL, and 1 L).
4. Inoculating loop.
5. Laminar airflow chamber.
6. White fluorescent lamps (2000–3000 lux).
7. pH meter.
8. Sterile distilled water.
9. Alcohol swabs.
10. Microscopic slides and cover slips.

3 Methods

1. Add 5 mL of pure *Chlorella* from a stock culture to 15 mL of BG11 N(+) culture media in a 50-mL conical flask. This mixture is grown under continuous aeration.
2. Insert an air filter at the head of a glass pipette and seal it with nonabsorbent cotton wool to serve as an aerating tube.
3. Allow the *Chlorella* to reach the stationary growth phase. Then transfer 50 mL of the culture into 200 mL of culture media in a 500-mL conical flask. Repeat the scale-up process to expand the culture [1].
4. Mix 300 mL of *Chlorella* culture with 2000 mL of culture medium in a Haffkine flask.
5. Estimate the growth rate of *Chlorella* sp. by measuring its optical density. Collect 5 mL of culture at 4-day intervals. During each sampling, 5 mL of freshly prepared sterile medium is added to correct for volume (Fig. 1).
6. Maintain culture flasks in a culture condition at (50 μmol photons/m2/s), 12 h light/dark, and 25 ± 2 °C for 30 days.
7. Measure algal growth using optical density. Record the OD of each culture sample at 600 nm with a UV-vis spectrophotometer (Optizen 3220; Japan).
8. Dilute samples when needed to keep OD_{600} values between 0.2 and 0.8 [2].

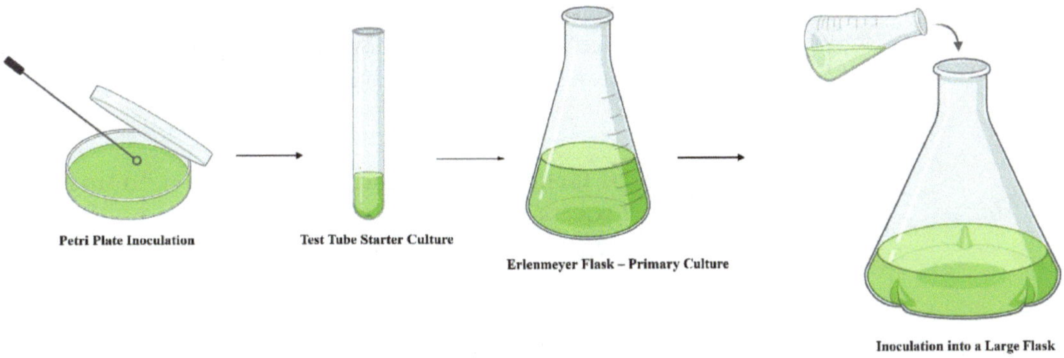

Fig. 1 Workflow for starter to large-scale *Chlorella* culture

9. Find biomass concentration by multiplying the OD_{600} value by 0.71, based on past data. Create a plot of OD_{600} versus dry cell weight (DCW) using samples with different biomass levels [3].

3.1 Observation

The *Chlorella* culture progressed from a light green to a dark green, indicating proper and continuous growth. During the cultivation phase, there are certainly no signs of contamination, indicating that aseptic handling is effective.

The optical density (OD_{600}) increased gradually with each measurement interval, indicating vigorous cell division and biomass development.

Aerated cultures grew more quickly and uniformly than stationary cultures, highlighting the need for sufficient aeration. Overall, biomass concentrations rise in accordance with OD_{600} values, indicating consistent and efficient culture development.

4 Notes

Proper aeration, light intensity, and temperature are critical for rapid and even microalgal growth. The culture remains contamination-free throughout the research, showing good aseptic handling. The starter culture has consistent growth patterns, making it ideal for subsequent scaling up, biomass generation, and biochemical analysis.

Data Availability Statement

All data generated or analyzed during this study are provided in the chapter and its supplementary information files.

Ethics Approval

Not applicable.

Competing Interests

The authors declare that they have no competing financial or nonfinancial interests that could have influenced the work reported in this chapter.

Declaration of Generative AI and AI-assisted Technologies in the Writing Process

During the preparation of this work, the author(s) used AI to improve the readability of the chapter. After using this tool, the author(s) reviewed and edited the content as needed and take(s) full responsibility for the content of the publication.

Funding

Department of Biotechnology (Govt. of India) for the establishment of the National Repository for Microalgae and Cyanobacteria—Freshwater & Marine (NRMC F & M) (Phase II) (BT/ PR29901/PBD26/694/2018) research. Chief Minister Research Grant (CMRG) for the financial support of research by the Government of Tamil Nadu, India, for the financial support (CMRG 2023–2024- Project No. CMRG2023BMBO7047).

References

1. Jui TJ, Tasnim A, Islam SR, Manjur OHB, Hossain MS, Tasnim N et al (2024) Optimal growth conditions to enhance Chlorella vulgaris biomass production in indoor phyto tank and quality assessment of feed and culture stock. Heliyon 10(11):e31826
2. Praveenkumar R, Shameera K, Mahalakshmi G, Akbarsha MA, Thajuddin N (2012) Influence of nutrient deprivations on lipid accumulation in a dominant indigenous microalga Chlorella sp., BUM11008: evaluation for biodiesel production. Biomass Bioenergy 37:60–66
3. Rajendran PJ, Shunmugam S, Saravanan C, Nooruddin T, Dharumadurai D (2025) Molecular monitoring, identification of the microcystin Synthetase (mcy) gene and Toxic potential of *Microcystis aeruginosa* in Indian freshwater blooms. Toxicon 253:108631

Chapter 20

Coculture of *Chlorella* and Bacteria for Enhanced Biomass Production

Asraf Sithikka Rasheed, Thajuddin Nooruddin, and Dhanasekaran Dharumadurai

Abstract

The strategic cocultivation approach of *Chlorella* species with bacteria renders a powerful approach of bioengineering to enhance the biomass productivity of microalgae, lipid accumulation, nutrient recover, and so on. This chapter offers a concise methodological framework for a model phycosphere that incorporates microalgae–bacteria consortium to optimize the yield of biomass and other compounds. This integrates the microalgae and bacterial strain selection criteria, inoculation ratios, and the core methodology that involves the axenic culture preparations, media optimization, growth conditions, monitoring parameters, harvesting methods, and the biomass characterization. This protocol presents a reproducible model for phycosphere-induced bioresource innovation by combining the ecological principles with cultivation strategies to obtain scalable biomass with desired biochemical outcomes.

Key words Cocultivation, *Chlorella*, Phycosphere, Microalgae, Bacteria consortium

1 Introduction

In aquatic environment, microalgae and heterotrophic bacteria are prevalent and coexist by depending on each other. Photosynthesis and microbial activity are positively correlated particularly in freshwater ecosystem [1]. Coculturing algae and bacteria in a system (Fig. 1) has been significantly gaining attention for its potential biomass enhancement and its dual benefits in increased biomass yield and wastewater treatment as certain bacteria can boost the microalgal growth by synergic interactions [2, 3]. The hybrid consortium with two different organisms is likely to adapt to different nutrient supplies and environmental conditions due to their varying metabolic activities [3]. In this consortium, the activity of aerobic bacteria mitigates the microalgal oxygen inhibition which in turn enhances the carbon assimilation in algae and stimulates the release of bacterial growth factors such as hormones and vitamins

Dhanasekaran Dharumadurai (ed.), *Chlorella Genomics, Cultivation, and Bioenterprise: Green Economy through Algal Innovations*, Springer Protocols Handbooks, https://doi.org/10.1007/978-1-0716-5276-3_20,
© The Author(s), under exclusive license to Springer Science+Business Media, LLC, part of Springer Nature 2026

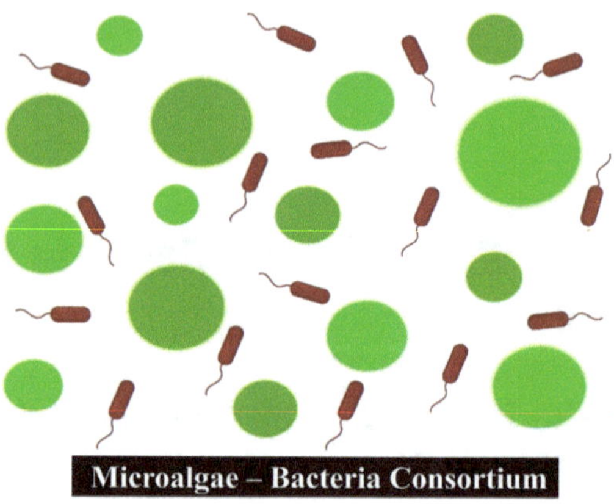

Fig. 1 Visual representation of coexistence of microalgae and bacteria—Phycosphere

[4]. Bacterial growth can be induced by algal dissolved organic carbon (DOC) typically at the end of the algal bloom with peak microbial activity which in turn regulates DOC levels, preventing its accumulation as observed in axenic cultures. These heterotrophic bacteria reside in a unique ecological niche known as phycosphere and frequently develop close association with live algae. It has been reported that recognized plant beneficial bacteria, such as Rhizobium, which also interact with algae, stimulate microalgal growth [1, 4, 5]. In the phycosphere, by the production of extracellular compounds like EPS, microalgae offer a protective environment that promotes bacterial growth. These metabolites play a vital role in the establishment of direct link between the two organisms, facilitating bacterial adhesion to the cell surface of microalgae. This relationship contributes the structural integrity and long term stability of the consortium during cultivation [1, 6, 7]. The mutual interaction between the microalgal–bacterial consortium has been schematically represented in Fig. 2. Beyond direct interaction, such plant beneficial bacteria can also remotely boost microalgal biomass and lipid content through gas exchange and volatile organic compounds. As the microalgal strains necessitates a range of nutrients such as nitrogen and phosphorus, the bacteria can mobilize and fix its nutritional requirements. This can be exemplified by coculturing *Chlorella pyrenoidosa* with nitrogen-fixing bacteria *Azotobacter beijerinckii* in wastewater, where the bacteria provide fixed nitrogen to the microalgae to promote its growth under nitrogen-limited conditions. This experiment resulted in 71.8% increase in the biomass compared to an axenic culture system, highlighting that bacterial nitrogen provision substantially enhances the biomass productivity and nutrient assimilation [2]. Furthermore, the coculturing system significantly enhanced the lipid yield. When *Chlorella sorokiniana* is cultivated with *Streptomyces thermocarboxydus*, it reported a

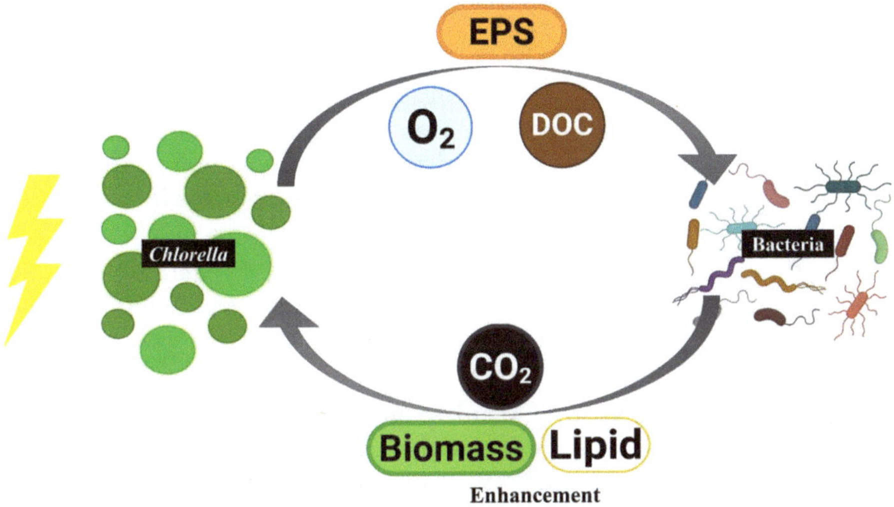

Fig. 2 Schematic illustration of mutual interactions between *Chlorella* and bacteria coculture

21–25% increase in both total biomass and lipid yield compared to its monoculture [8]. This chapter provides a comprehensive step by step protocol for coculturing *Chlorella* and bacteria together as a phycosphere model for enhanced biomass production.

2 Materials

1. Separate axenic cultures of both microalgae and bacteria.
2. Sterile growth media.
3. Analytical reagents (e.g., Solvents such as 90% acetone or methanol for chlorophyll extraction).
4. Biosafety cabinet.
5. Autoclave.
6. Incubator shaker and orbital shaker.
7. Illuminated incubator.
8. Erlenmeyer shake flasks/photobioreactors/biofilm cultivation plates.
9. While LEDs or fluorescent lamps.
10. Air pumps with sterile air filters.
11. UV–Vis spectrophotometer.
12. Light microscope.
13. pH meter.
14. Centrifuge/filtration setups/flocculants.
15. Basic glasswares and pipettes.

3 Methods

1. Select the microalgal strain—*Chlorella* species according to the target application such as wastewater bioremediation, lipid, or biomass yield and so on. The most prevalently utilized strains include *Chlorella vulgaris*, *Chlorella sorokiniana*, and *Chlorella pyrenoidosa*.
2. Select the bacterial partners that facilitates the mutual interaction with the microalgae thereby enhancing the yield such as *Stenotrophomonas maltophilia*, *Rhizobium radiobacter*, and *Azotobacter beijerinckii*.
3. Prepare a standard growth medium for microalgae with all essential micro and macro nutrients such as BG 11+. Add low concentrations of organic carbon source while incorporating heterotrophic or mixotrophic bacterial partners to avoid bacterial overgrowth. Adjust the nitrogen sources carefully as the nitrogen-fixing bacteria can remodel the nitrogen profile in the medium. This allows for sufficient nitrogen supply while avoiding additional nitrogen supplementation.
4. Culture the *Chlorella* in sterile BG 11+ medium under controlled laboratory conditions with the light intensity of ~100–200 μmol photons m^{-2} s^{-1} and the temperature about 25 °C (*see* **Notes 1–3**).
5. Maintain the cultures in log phase for 3–5 days and determine the cell density through direct cell counting through microscope or by optical density at 680 nm to plan subsequent inoculation for coculture.
6. Isolate and culture the bacterial strains in sterile nutrient or LB broth and wash the harvested bacterial cells to remove the residual nutrients and standardize it to attain the optical density at 600 nm.
7. Choose an initial bacterium to algae inoculation ratio as this parameter critically determines the outcome of the consortium (Employ a 1:1 ratio to achieve a balanced, mutual nutrient exchange as a standard initial phase; use a 10:1 ratio for enhanced bioremediation).
8. Choose a cultivation system based on the production scale, Shake flasks for lab-scale screening, bench top photobioreactors for pilot scale screening, and biofilm plates or attached systems for the cocultures that requires close association.
9. Maintain a consistent temperature about 20–30 °C and 100–200 μmol m^{-2} s^{-1} illumination with the photoperiod of 12:12 or 16:8 h (light:dark) cycle for the cultivation (*see* **Note 5**).

10. Monitor the growth of the culture through cell density, chlorophyll estimation, and assessing the pH and dissolved oxygen (*see* **Note 4**).
11. After the growth period (typically in 8–14 days), harvest the biomass through centrifugation, flocculation, or filtration.
12. Quantify and qualify the biomass by their dry weight and lipid, protein, or carbohydrate evaluation for the desired applications [2, 9–12].

4 Notes

1. Maintain axenic initial cultures for both organisms to avoid contamination and to prevent nutrient remodeling and for population balance.
2. Employ aseptic techniques throughout the experiment in sterile environment.
3. Avoid excessive organic carbon supplementation which may lead to bacterial overgrowth, oxygen depletion, and algal bleach.
4. Consistently monitor the cultures for purity, pH, salinity, and dissolved oxygen.
5. Maintain the illumination, temperature, and aeration standard.

Acknowledgments

We sincerely acknowledge the financial support from the DBT (Govt. of India), for the establishment of National Repository for Microalgae and Cyanobacteria—Freshwater & Marine (NRMC-F & M) (Phase II) (BT/PR29901/PBD26/694/2018). We gratefully acknowledge the Chief Minister Research Grant (CMRG), Govt. of Tamil Nadu for providing financial support (CMRG 2023-2024 Project No. CMRG2023BMBO7047), Rashtriya Uchchatar Shiksha Abhiyan (RUSA) – 2.0 Biological Sciences (TN RUSA: 311/RUSA (2.0)/2018 Dt. December 02, 2020) and School of Life Sciences, Bharathidasan University, Tiruchirappalli, Tamil Nadu, India.

References

1. Berthold DE, Shetty KG, Jayachandran K, Laughinghouse HD IV, Gantar M (2019) Enhancing algal biomass and lipid production through bacterial co-culture. Biomass Bioenergy 122:280–289
2. Dong H, Liu W, Zhang H, Wang Z, Feng F, Zhou L, Duan H, Xu T, Li X, Ma J (2023) Enhanced biomass production and wastewater treatment in attached co-culture of Chlorella pyrenoidosa with nitrogen-fixing bacteria

Azotobacter beijerinckii. Bioprocess Biosyst Eng 46(5):707–716

3. Boonma S, Chaiklangmuang S, Chaiwongsar S, Pekkoh J, Pumas C, Ungsethaphand T, Tongsiri S, Peerapornpisal Y (2015) Enhanced carbon dioxide fixation and bio-oil production of a microalgal consortium. CLEAN Soil, Air, Water 43(5):761–766
4. Amavizca E, Bashan Y, Ryu CM, Farag MA, Bebout BM, de-Bashan LE (2017) Enhanced performance of the microalga Chlorella sorokiniana remotely induced by the plant growth-promoting bacteria Azospirillum brasilense and Bacillus pumilus. Sci Rep 7(1):41310
5. Kim BH, Ramanan R, Cho DH, Oh HM, Kim HS (2014) Role of rhizobium, a plant growth promoting bacterium, in enhancing algal biomass through mutualistic interaction. Biomass Bioenergy 69:95–105
6. Huo S, Kong M, Zhu F, Qian J, Huang D, Chen P, Ruan R (2020) Co-culture of Chlorella and wastewater-borne bacteria in vinegar production wastewater: enhancement of nutrients removal and influence of algal biomass generation. Algal Res 45:101744
7. Gonçalves AL, Pires JC, Simões M (2017) A review on the use of microalgal consortia for wastewater treatment. Algal Res 24:403–415
8. Padri M, Boontian N, Teaumroong N, Piromyou P, Piasai C (2022) Co-culture of microalga Chlorella sorokiniana with syntrophic Streptomyces thermocarboxydus in cassava wastewater for wastewater treatment and biodiesel production. Bioresour Technol 347:126732
9. Gopalakrishnan K, Wager YZ, Roostaei J (2024) Co-cultivation of microalgae and bacteria for optimal bioenergy feedstock production in wastewater by using response surface methodology. Sci Rep 14(1):20703
10. Xue L, Shang H, Ma P, Wang X, He X, Niu J, Wu J (2018) Analysis of growth and lipid production characteristics of Chlorella vulgaris in artificially constructed consortia with symbiotic bacteria. J Basic Microbiol 58(4):358–367
11. Tavakol M, Naeimpoor F (2025) Single and co-cultivation of Chlorella vulgaris and Pseudomonas aeruginosa: efficient nutrient removal and lipid/starch formation by photoperiodic co-culture. Bioresour Technol 436:132983
12. Xu M, Xue Z, Sun S, Zhao C, Liu J, Liu J, Zhao Y (2020) Co-culturing microalgae with endophytic bacteria increases nutrient removal efficiency for biogas purification. Bioresour Technol 314:123766

Chapter 21

Scaling up of *Chlorella* spp. from Lab to Pilot Scale

Fayaazuddin Thajuddin, Thajuddin Nooruddin, and Dhanasekaran Dharumadurai

Abstract

Microalgae, especially *Chlorella* species, are regarded as essential photosynthetic cell factories for the production of high-value-added products and biodiesel. Present protocol outlines a step-by-step procedure from lab-scale maintenance to pilot-scale production in transparent plastic containers with nutrient supplementation and offers a comprehensive, optimized strategy for scaling up *Chlorella* spp. Throughout the scale-up process, the protocol maintains ideal growth conditions for increased production of algal biomass, thus, offering a dependable route to sustained microalgal production.

Key words *Chlorella*, Algal biomass, Biodiesel, Lab-scale cultivation, Pilot-scale production, Scale-up process optimization

1 Introduction

Microalgae are photoautotrophic microorganisms that significantly transform carbon dioxide into valuable bioactive compounds [1]. They possess a wide range of incomparable advantages such as simple cultivation methods, rapid productivity, and adaptability [2, 9]. With remarkable growth patterns and photosynthetic capacity, microalgae have the most straightforward mechanism in nature. *Chlorella* is a single-celled eukaryotic green algae present in both fresh and marine environments. It can be cultivated under adaptable environments including autotrophic, heterotrophic, mixotrophic, or photoheterotrophic conditions [3]. Their enriched nutrient profile including high-quality vitamins, proteins, lipids, and other valuable compounds makes them a promising candidate to be used as food additives [4]. Biofuels derived from algae such as biodiesel possess promising biochemical profile to be utilized as an alternative for conventional fuel. Additionally, oleaginous microalgae present a distinct advantage as feedstock for biofuel production when compared to food crops. Besides, their short-term growth

Dhanasekaran Dharumadurai (ed.), *Chlorella Genomics, Cultivation, and Bioenterprise: Green Economy through Algal Innovations*, Springer Protocols Handbooks, https://doi.org/10.1007/978-1-0716-5276-3_21,
© The Author(s), under exclusive license to Springer Science+Business Media, LLC, part of Springer Nature 2026

cycle and their ability to be grown in wastewater and abandoned land can possibly lead to yield increased lipid content. *Chlorella protothecoides* exemplifies the enhanced lipid yield for about four-fold under heterotrophic conditions than under normal photoautotrophic conditions [6]. Additionally, the heterotrophic cultivation of *Chlorella zofingiensis* resulted in high levels of neutral lipids, while the photoautotrophic cultivation favored the enhanced accumulation of phospholipid and glycolipids. Thus, remarkably under various growth conditions, both the quantity and quality of the lipids considerably fluctuate [7].

Hence, utilizing microalgae has immense potential as raw material for biodiesel production and can avoid competition of limited arable land and freshwater [5]. Lastly, high culture medium cost and intricate processing pose a significant bottleneck for the commercialization of algal biomass. Therefore, the present chapter provides a detailed and optimized protocol on scaling up of *Chlorella* spp., from laboratory conditions to pilot scale production for higher algal biomass production.

2 Materials

1. *Chlorella* spp.
2. BG 11 (N^+) medium.
3. 250 mL and 3000 mL conical flask.
4. 25 L plastic container.
5. Ethanol.
6. UV light.
7. Autoclave.

2.1 Composition of *Chlorella spp.* Growth Medium

2.1.1 BG—11 (N^+) Medium (g/L) [8]

Chemicals	g/L
Sodium nitrate	1.5
Dipotassium hydrogen phosphate	0.04
Magnesium sulphate	0.075
Calcium chloride	0.036
Citric acid	0.006
Ferric ammonium citrate	0.006
EDTA (disodium salt)	0.001

(continued)

Chemicals	g/L
Sodium carbonate	0.02
Trace metal mix A5	1.0 ml
Distilled water	1.0 L

3 Methods

3.1 Collection of Microalgae Culture and Maintenance

1. Microalgae strain *Chlorella* spp., is acquired from microalgae and cyanobacterial culture collection center.
2. Inoculate the *Chlorella* spp., in a conical flask with a capacity of 250 mL and working volume of 100 mL.
3. Incubate the cultures in sterile BG-11(N^+) growth medium and maintain under white color fluorescent lamps with 50 μmol photons/m^2/s 12 h light/dark cycle at 25 ± 2 °C for 30 days (*see* **Note 1**) [5].
4. To prevent cell clumping, the cultures are subjected to manual shaking every day.

3.2 Scaling up of Microalgae in Laboratory Scale

1. Inoculate the 100 mL of culture in the 900 mL of BG-11(N^+) growth medium (1:2 ratio) in sterile 3000 mL conical flask.
2. Cover the mouth of conical flask with sterilized cotton plug.
3. Allow the microalgae cells to multiply for 10–15 days under optimized culture conditions.
4. Later, transfer 1000 mL of medium to make up the final volume of 2000 mL and incubate the same conditions for 20 days.

3.3 Scaling up of Microalgae to Pilot Scale

1. Sterilize 25 L plastic container using ethanol and place it under UV light for 15 min.
2. After sterilization, inoculate 2000 mL of *Chlorella* spp., in 4000 mL of BG-11(N^+) growth medium in the plastic container.
3. Incubate the 25 L container under proper light illumination and aeration to avoid cell clumping (*see* **Notes 2–3**).
4. Then, every 10 days add 1000 mL of growth medium until it reaches the final volume of 20 L.
5. Repeat the same procedure to prepare multiple plastic container with *Chlorella* spp., for pilot scale cultivation (Fig. 1).

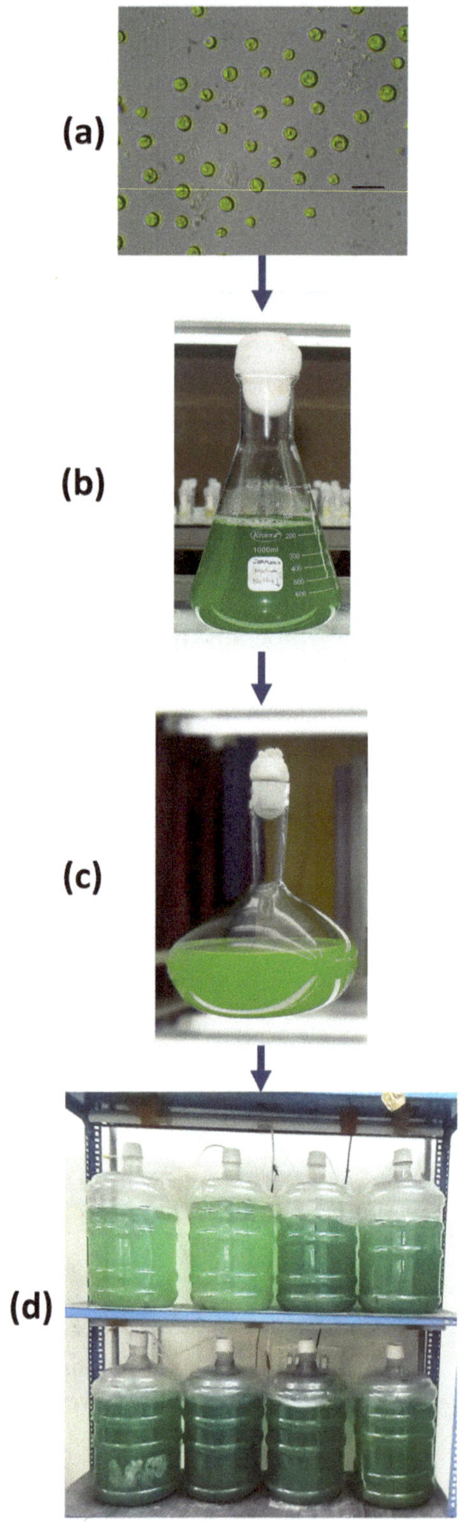

Fig. 1 Graphical illustration of *Chlorella* spp., from laboratory to pilot scale—(**a**) *Chlorella* spp., (**b**) 1000 mL conical flask, (**c**) 3000 mL conical flask, (**d**) 25 L plastic container

4 Notes

1. For optimal growth, the BG-11(N^+) medium requires precise preparation and sterilization.
2. Twenty-five liter plastic cans must be transparent to ensure proper light illumination for microalgae.
3. Aeration tubes must be sterilized and cleaned regularly.

Acknowledgments

We sincerely acknowledge the financial support from the DBT (Govt. of India), for the establishment of National Repository for Microalgae and Cyanobacteria—Freshwater & Marine (NRMC-F & M) (Phase II) (BT/PR29901/PBD26/694/2018). We gratefully acknowledge the Chief Minister Research Grant (CMRG), Govt. of Tamil Nadu for providing financial support (CMRG 2023-2024 Project No. CMRG2023BMBO7047), Rashtriya Uchchatar Shiksha Abhiyan (RUSA) – 2.0 Biological Sciences (TN RUSA: 311/RUSA (2.0)/2018 Dt. December 02, 2020) and Department of Microbiology, Bharathidasan University, Tiruchirappalli, Tamil Nadu, India.

References

1. Chisti Y (2007) Biodiesel from microalgae. Biotechnol Adv 25(3):294–306
2. Potvin G, Zhang Z (2010) Strategies for high-level recombinant protein expression in transgenic microalgae: a review. Biotechnol Adv 28(6):910–918
3. Mata TM, Martins AA, Caetano NS (2010) Microalgae for biodiesel production and other applications: a review. Renew Sust Energ Rev 14(1):217–232
4. Wang SK, Hu YR, Wang F, Stiles AR, Liu CZ (2014) Scale-up cultivation of Chlorella ellipsoidea from indoor to outdoor in bubble column bioreactors. Bioresour Technol 156:117–122
5. Arutselvan C, Narchonai G, Pugazhendhi A, LewisOscar F, Thajuddin N (2021) Evaluation of microalgal strains and microalgal consortium for higher lipid productivity and rich fatty acid profile towards sustainable biodiesel production. Bioresour Technol 339:125524
6. Xu H, Miao X, Wu Q (2006) High quality biodiesel production from a microalga Chlorella protothecoides by heterotrophic growth in fermenters. J Biotechnol 126(4):499–507
7. Liu J, Huang J, Sun Z, Zhong Y, Jiang Y, Chen F (2011) Differential lipid and fatty acid profiles of photoautotrophic and heterotrophic Chlorella zofingiensis: assessment of algal oils for biodiesel production. Bioresour Technol 102(1):106–110
8. Rippka R, Deruelles J, Waterbury JB, Herdman M, Stanier RY (1979) Generic assignments, strain histories and properties of pure cultures of cyanobacteria. Microbiology 111(1):1–61
9. Thajuddin F, Rasheed AS, Elumalai A, Palanivel P, Nooruddin T, Dharumadurai D (2026) Harnessing microalgae and cyanobacteria for sustainable bioenergy production: integration of molecular engineering and AI-driven approaches. Renew Energy 256:123961

Chapter 22

Large-Scale Cultivation of *Chlorella* in Open Pond Systems

Fayaazuddin Thajuddin, Thajuddin Nooruddin,
and Dhanasekaran Dharumadurai

Abstract

Microalgae are a crucial resource for biofuels, nutraceuticals, and animal feed due to the growing demand for sustainable biomass worldwide. The scalability and affordability of open pond systems, especially raceway ponds, make them popular for large-scale cultivation. For the large-scale production of *Chlorella* spp. employing open pond and raceway pond systems, this chapter offers a comprehensive and standardized approach. In order to produce high algal biomass, it describes the necessary materials, the composition of the growth medium, and the detailed methods for growing and harvesting. To guarantee efficient and repeatable manufacturing, crucial operational factors comprising controlling contamination, monitoring pH and temperature, and regulating paddle wheel speed are highlighted. The purpose of this protocol is to make commercial production of *Chlorella* for a variety of biotechnological applications.

Key words Microalgae, *Chlorella*, Outdoor Pond, Raceway Pond, Biomass production, Harvesting methods

1 Introduction

Microalgae are promising feedstock for a wide range of applications such as human food supplement, animal feed, fish feed, medicine, and bioenergy [1, 2]. Outdoor cultivation systems are standard for the commercial production of microalgae as they provide a simple and straightforward method for the large-scale production of algal biomass, where the closed systems are used less common. The open cultivation systems completely depend on the natural environmental conditions, for example, high temperature causes evaporation [3]. Open raceway pond was considered as primary option for the cost-effective cultivation of large-scale microalgal biomass since 1950s. A raceway pond is constructed as a closed-loop circulating channel with 0.3 m depth for the algal circulation. The entire channel is built of the concrete wall with 3.95 m thickness along with 0.1 m thick wall of central divider [4, 5]. The construction of these ponds is shallow and designed to facilitate the culture

Dhanasekaran Dharumadurai (ed.), *Chlorella Genomics, Cultivation, and Bioenterprise: Green Economy through Algal Innovations*, Springer Protocols Handbooks, https://doi.org/10.1007/978-1-0716-5276-3_22,
© The Author(s), under exclusive license to Springer Science+Business Media, LLC, part of Springer Nature 2026

circulation and paddle wheels in the center of the pond mixes the nutrients in the media, air, and the cells evenly throughout the pond [6]. The main disadvantage of the open raceway pond cultivation is the long light path and the ability of the culture to get contaminated by various external factors such as dust materials, birds, and microorganisms [7]. For a standard circulation, the flow velocity must be maintained between 0.10 and 0.24 m/s, which is generally achieved by the paddle wheel rotated at 35–55 rpm [5, 10]. The algal biomass cultivated through the raceway ponds are typically used for animal feed and biodiesel production. Thus, the present chapter provides a detailed protocol on large-scale of cultivation of *Chlorella* spp., in open pond and raceway pond system for higher algal biomass cultivation.

2 Materials

1. *Chlorella* spp.
2. 30 L open pond.
3. 5 KL and 35 KL open raceway pond.
4. Urea.
5. Superphosphate.
6. Plankton net.
7. UV-vis spectrophotometer.
8. Thermometer.
9. pH meter.
10. Water.

2.1 Composition of Growth Medium

Outdoor pond cultivation medium (mg/L)

Chemicals	mg/L
Urea —(CH_4N_2O)	100
Superphosphate — $Ca(H_2PO_4)_2$	80

3 Methods

3.1 Outdoor pond Cultivation of Chlorella spp.

1. Collect the microalgae strain *Chlorella* spp., from the culture collection center.

2. Prepare outdoor pond with total volume of 45 L, 82 cm length, 28 cm depth, and working volume of 30 L.
3. Clean the pond thoroughly to remove any contamination.
4. Measure and mix the cultivation medium with urea (CH_4N_2O) and superphosphate ($Ca(H_2PO_4)_2$) in a beaker using weighing balance for 30 L.
5. Now, transfer the prepared medium and inoculate 10% (v/v) of *Chlorella* spp., in the outdoor pond.
6. Maintain the final density of initial inoculum at 0.2 OD using UV-vis spectrophotometer.
7. Regularly, check the temperature, pH, and cell density (*see* **Notes 1–3**).
8. Transfer or harvest *Chlorella* spp., from the pond, when the final cell density reaches more than 2.2 OD [9].

3.2 Open Raceway pond Cultivation of Chlorella *spp.*

1. Initially in the seed raceway pond, prepare the medium for 5 KL (5000 L) with *Chlorella* spp., initial inoculum of 10% at 0.2 OD.
2. Maintain the paddle wheel speed at 15 rpm for assisting in continuous flow throughout the raceway pond.
3. Secondly, when the cell density reached at 2.2 OD in 5 KL raceway pond. It serves as an inoculum for 35 KL raceway pond.
4. Now, transfer the *Chlorella* spp., biomass to 35 KL raceway pond and add the nutritional growth accordingly.
5. Monitor the temperature, pH, and cell density regularly (*see* **Notes 1–3**).
6. *Chlorella* spp. can be harvested when the final cell density reaches more than 2.2 OD (Fig. 1).
7. For continuous cultivation of *Chlorella* spp., after harvesting, replenish the raceway pond with water and growth medium for every 2 days period [8].

3.3 Harvesting and Drying of Chlorella *spp.*

1. Prepare plankton filter net with a pore size of 50 μm for harvesting.
2. Attach the plankton net in a rectangular stainless-steel frame for the *Chlorella* spp., wet biomass to be harvested efficiently [8].
3. Collect the wet biomass from the plankton net with 5000 mL beaker and allow it to dry for 24–30 h.
4. Carry out the process in sun or shade drying method (Fig. 2).
8. Dried biomass is stored at room temperature for other experimental or commercial purpose.

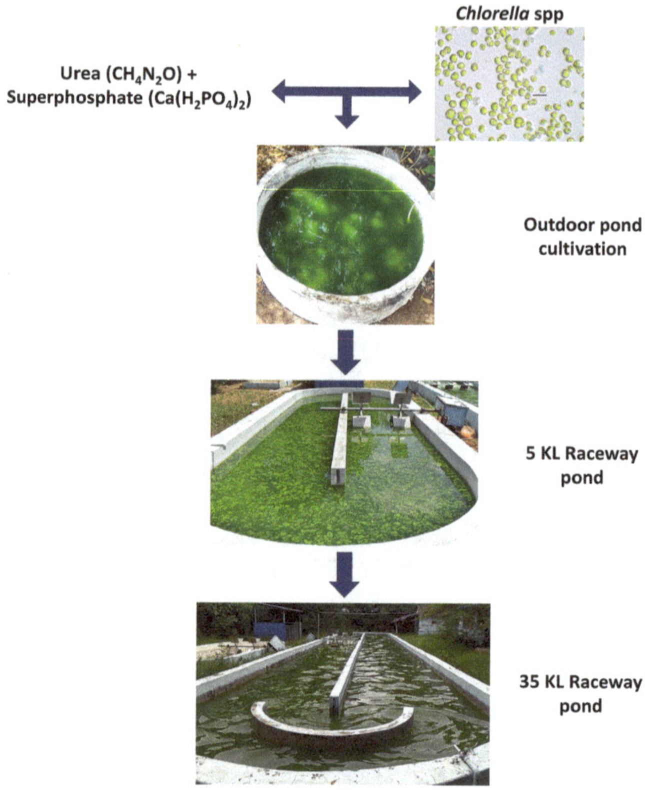

Fig. 1 Step-by-step cultivation of *Chlorella* using Open Pond systems

Fig. 2 Downstream processing of *Chlorella* biomass

4 Notes

1. Regularly, remove the leaves and other environmental waste from the open raceway pond using fish/mosquito net.
2. Water loss occurs due to evaporation. Therefore, check the water level periodically and replenish same quantity of water with growth medium.
3. Periodically, harvest the algal biomass to prevent cell death.

Acknowledgments

We sincerely acknowledge the financial support from the DBT (Govt. of India), for the establishment of National Repository for Microalgae and Cyanobacteria—Freshwater & Marine (NRMC-F & M) (Phase II) (BT/PR29901/PBD26/694/2018). We gratefully acknowledge the Chief Minister Research Grant (CMRG), Govt. of Tamil Nadu for providing financial support (CMRG 2023-2024 Project No. CMRG2023BMBO7047), Rashtriya Uchchatar Shiksha Abhiyan (RUSA) – 2.0 Biological Sciences (TN RUSA: 311/RUSA (2.0)/2018 Dt. December 02, 2020) and Department of Microbiology, Bharathidasan University, Tiruchirappalli, Tamil Nadu, India.

References

1. Harun R, Singh M, Forde GM, Danquah MK (2010) Bioprocess engineering of microalgae to produce a variety of consumer products. Renew Sust Energ Rev 14(3):1037–1047
2. Thajuddin F, Rasheed AS, Elumalai A, Palanivel P, Nooruddin T, Dharumadurai D (2026) Harnessing microalgae and cyanobacteria for sustainable bioenergy production: integration of molecular engineering and AI-driven approaches. Renew Energy 256: 123961
3. Tredici MR (2004) Mass production of microalgae: photobioreactors. In: Handbook of microalgal culture: Biotechnology and applied phycology, vol 1, pp 178–214
4. Sompech K, Chisti Y, Srinophakun T (2012) Design of raceway ponds for producing microalgae. Biofuels 3(4):387–397
5. Borowitzka MA (1999) Commercial production of microalgae: ponds, tanks, tubes and fermenters. J Biotechnol 70(1–3):313–321
6. Arutselvan C, Kumar Seenivasan H, Oscar FL, Ramya G, Chi NTL, Pugazhendhi A, Thajuddin N (2022) Review on wastewater treatment by microalgae in different cultivation systems and its importance in biodiesel production. Fuel 324:124623
7. Chisti Y (2013) Raceways-based production of algal crude oil. Green 3(3–4):195
8. Baldev E, Mubarakali D, Saravanakumar K, Arutselvan C, Alharbi NS, Alharbi SA, Sivasubramanian V, Thajuddin N (2018) Unveiling algal cultivation using raceway ponds for biodiesel production and its quality assessment. Renew Energy 123:486–498
9. Arutselvan C, Narchonai G, Pugazhendhi A, LewisOscar F, Thajuddin N (2021) Evaluation of microalgal strains and microalgal consortium for higher lipid productivity and rich fatty acid profile towards sustainable biodiesel production. Bioresour Technol 339:125524
10. Fayaazuddin T, Prakash P, Shakena Fathima T, Dhanasekaran D (2023) Commercial astaxanthin production from green alga Haematococcus pluvialis. In: Food microbiology based entrepreneurship: making money from microbes. Springer, Singapore, pp 279–304

Chapter 23

Harvesting Techniques of *Chlorella* Biomass Using Filtration and Sedimentation

Prakash Palanivel, Priyanka Jayam Rajendran, Fayaazuddin Thajuddin, Dhanasekaran Dharumadurai, and Gangatharan Muralitharan

Abstract

Chlorella is well-known for its biotechnological potential due to its ability to grow quickly, adapt to a wide range of environmental circumstances, and create useful proteins, lipids, pigments, and bioactive chemicals. Nevertheless, biomass gathering continues. However, biomass harvesting remains problematic due to the small cell size, low density, and capacity to create a stable suspension in water. This chapter describes the procedures for filtering and sedimentation, two standard laboratory-scale harvesting methods, to evaluate the effectiveness with which they harvest biomass. While filtering provides for fast biomass recovery with vacuum support, sedimentation is a simple and cost-effective approach in which cells naturally settle due to gravity. When these methods are combined, they provide important information on how to select an efficient method for harvesting *Chlorella* in a laboratory setting.

Key words *Chlorella*, Biomass harvesting, Sedimentation, Filtration, Biomass recovery, Microalgal cultivation

1 Introduction

Chlorella has received a lot of interest as a sustainable source of proteins, lipids, pigments, and bioactive compounds. They are suitable for large-scale production because they can flourish in a range of settings, grow quickly, and use sunlight efficiently [1]. Harvesting microalgal biomass, however, remains extremely difficult. The method is challenging due to the small cell size (2–10 μm), low cell density, and stable suspension in water. Harvesting, which accounts for 20–30% of overall production costs, is a key stage in downstream processing [2].

Centrifugation, flocculation, filtration, and sedimentation are all methods used to collect biomass [3]. Sedimentation and filtration are the simplest and most affordable procedures. Sedimentation uses gravity to separate algal cells from their growth media,

Dhanasekaran Dharumadurai (ed.), *Chlorella Genomics, Cultivation, and Bioenterprise: Green Economy through Algal Innovations*, Springer Protocols Handbooks, https://doi.org/10.1007/978-1-0716-5276-3_23,
© The Author(s), under exclusive license to Springer Science+Business Media, LLC, part of Springer Nature 2026

depending on density differences. This process is slow, yet inexpensive and energy-efficient. Filtration, on the other hand, uses filter paper or porous membranes to separate microalgae from the liquid medium. This technology allows for faster separation, but it is restricted by filter clogging and requires extensive maintenance in large systems. Filtration and sedimentation are both effective methods for small-volume cultures in the laboratory. Sedimentation is successful when flocculation occurs naturally or in dense populations. Filtration is appropriate for investigations that require rapid, clean biomass recovery [4].

Understanding the efficacy of these two strategies assists in choosing the best harvesting strategy for *Chlorella* biomass. In this experiment, *Chlorella* cultures are harvested using sedimentation and filtration methods. The effectiveness of separation, clarity of the supernatant, convenience of collection, and visual appearance of the recovered biomass are evaluated. The findings shed light on the practicality and efficacy of each procedure under laboratory circumstances. This chapter describes a protocol for obtaining *Chlorella* biomass using sedimentation and filtering methods. We assess recovered biomass appearance, ease of collection, supernatant purity, and separation efficiency. This helps us learn how beneficial and efficient each strategy is in laboratory environments.

2 Materials

1. Fresh *Chlorella* culture (log phase).
2. Conical flasks or glass beakers (500 mL).
3. Filter paper (Whatman No. 1 or equivalent).
4. Vacuum filtration unit or funnel.
5. Centrifuge tubes.
6. Measuring cylinder.
7. Distilled water.
8. Drying oven or desiccator.
9. Analytical balance.

3 Methodology

3.1 Sedimentation Technique

1. Transfer equal volumes (e.g., 250 mL) of *Chlorella* culture to sterile conical flasks.
2. Allow the flasks to remain undisturbed under diffused light at room temperature.
3. Monitor the culture at regular intervals (1 h, 6 h, 12 h, and 24 h).

4. Take note of any noticeable changes in color intensity and the separation of the clear liquid from the solid material.
5. After 24 h, carefully remove the clear liquid without disturbing the settled biomass.
6. Collect the settled biomass to dry and weigh.

3.2 Filtration Technique

1. Use a fresh Whatman No. 1 filter paper to assemble the vacuum filtration system.
2. Pour the *Chlorella* culture onto the filter and apply gentle vacuum suction.
3. Keep filtering until most of the liquid is filtered through the filter paper.
4. Gently wash the retained biomass with distilled water to remove any leftover media.
5. Use a spatula to scrape the wet biomass from the filter paper.
6. Transfer collected biomass to a pre-weighed plate and dry in a hot air oven (60 °C) until appropriate weight is achieved.
7. Record dried the weight of the dried biomass mass.

3.3 Expected Observations

The research study was completed successfully utilizing sedimentation and filtering techniques (Fig. 1). The green color of the supernatant disappeared during sedimentation. After 24 h, a thick layer of dense green *Chlorella* biomass accumulated at the bottom of the flask, but the upper layer remained clear and transparent. This demonstrated that the algae cells were effectively settled under gravitational force.

In the filtration method, the culture passed through the filter paper with increasing resistance over time. The green layer on the filter surface thickened and the filtrate became more transparent.

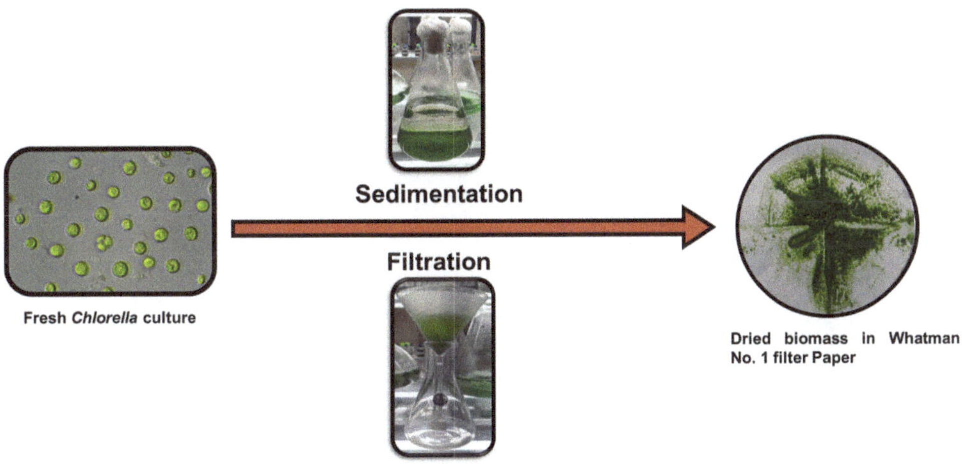

Fig. 1 Schematic representation of Chlorrella cells, biomass harvesting, and concentrated biomass collection

The biomass obtained was dense, moist, and simple to scrape off. Even while some filter clogging occurred later, hoover filtering was faster than sedimentation. The filtration-dried biomass appears cleaner and more homogenous in texture, but the sedimented sample has some remaining medium and is slightly thicker.

4 Notes

1. Clearly identify flasks, filter sheets, and drying plates with sample ID, date, and technique.
2. Monitor sedimentation levels to ensure *Chlorella* cells are separated by gravity.
3. Track the color, clarity, and biomass accumulation during deposition. Ensure that the vacuum filtration device is correctly assembled during the filtration process.
4. To keep liquids flowing smoothly and biomass retained, replace clogged filter papers as needed.
5. At 60 °C, all recovered biomass must be dried.
6. To use an analytical balance to accurately assess the biomass's yield, appearance, and purity, keep a thorough track of the drying weights and biomass recovery.

5. References

1. Mata TM, Martins AA, Caetano NS (2010) Microalgae for biodiesel production and other applications: a review. Renew Sust Energ Rev 14(1):217–232
2. Uduman N, Qi Y, Danquah MK, Forde GM, Hoadley A (2010) Dewatering of microalgal cultures: a major bottleneck to algae-based fuels. J Renew Sustain Energy 2(1):012701
3. Christenson L, Sims R (2011) Production and harvesting of microalgae for wastewater treatment, biofuels, and bioproducts. Biotechnol Adv 29(6):686–702
4. Addy MM, Zhang R, Ruan R (2016) Harvesting and processing of microalgal biomass for biofuel production. In: Biofuels from algae. Elsevier, pp 265–284

Chapter 24

Post-harvest Processing of *Chlorella* Biomass: Drying and Powdering

Prakash Palanivel, Gokul Varutharaj, Dhanasekaran Dharumadurai, and Gangatharan Muralitharan

Abstract

The post-harvest treatment of *Chlorella* biomass determines its stability, long-term storage, and suitability for downstream applications such as biofuels, pigments, nutritional supplements, and animal feed. Due to its high-water content, freshly harvested *Chlorella* is subject to microbial deterioration and is difficult to store. This chapter describes a typical method for drying and powdering *Chlorella* biomass using classic techniques such as freeze-drying, oven-drying, and sun- or solar-tunnel drying. The procedure determines the optimal drying conditions, reduction of moisture targets, and basic handling methods for preserving the structural integrity of critical components such as pigments, proteins, and lipids. A specialized powdering process is also incorporated, which produces consistent, free-flowing microalgal powder suitable for storage or processing. This technique thus provides a helpful guideline for corporations and researchers who want frequent, high-quality post-harvest processing of *Chlorella* biomass.

Key words *Chlorella*, Post-harvest processing, Drying, Powdering, Milling, Biomass preservation, Algal powder

1 Introduction

Chlorella are among the most promising biological resources. They have high photosynthesis efficiency, develop quickly, and contain valuable biological compounds. *Chlorella* biomass is used in many different applications, including biofuels, nutraceuticals, medicines, pigments, food additives, and aquaculture feed [1]. However, freshly obtained *Chlorella* contains 80 to 90% water. This significantly destabilises the biomass; hence, early treatment after harvest is critical. If not properly dried, biomass is susceptible to microbial contamination, enzymatic degradation, and the rapid loss of pigments, lipids, and proteins [2, 3].

Drying represents an important procedure in maintaining biomass stability, reducing storage capacity, and preparing the material

Dhanasekaran Dharumadurai (ed.), *Chlorella Genomics, Cultivation, and Bioenterprise: Green Economy through Algal Innovations*, Springer Protocols Handbooks, https://doi.org/10.1007/978-1-0716-5276-3_24,
© The Author(s), under exclusive license to Springer Science+Business Media, LLC, part of Springer Nature 2026

for eventual extraction or formulation. Various drying methods have been developed based on the desired product, each with its own set of advantages and disadvantages. Sun-drying provides cost-effective solutions for large industrial applications, although changing climatic conditions may induce pigment degradation [4]. Oven-drying is commonly employed in laboratory procedures because it provides controlled temperature conditions. High temperatures, on the other hand, can degrade thermally sensitive molecules such as chlorophyll and carotenoids [5]. Although more expensive and energy-intensive, freeze-drying has been proven to better retain biological constituents such as proteins, antioxidants, and vital pigments [1]. A standardized, step-by-step procedure for drying and powdering is critical for assuring reproducibility, maintaining quality, and facilitating useful cross study comparisons. This chapter provides a scientifically validated approach to post-harvest processing of *Chlorella* biomass that contains insights into drying kinetics, biochemical stability, and industrial research (Fig. 1).

2 Materials

1. Harvested wet *Chlorella* biomass.
2. Drying apparatus depending on chosen method—sun-drying/laboratory oven/freeze-dryer (Lyophilizer).
3. Clean cloth.
4. Sand Bed with mesh.
5. Desiccator.
6. Grinding mill.
7. Analytical balance.
8. Thermogravimetric analyzer.

3 Methods

1. Harvest *Chlorella* culture using centrifugation, filtration or flocculation.
2. Decant the supernatant, followed by a low-speed spin to remove excess water.
3. Concentrate the biomass to obtain a paste or slurry with approximately 20–30% solid content before drying.
4. Use an analytical balance to determine the moist weight of the concentrated biomass.

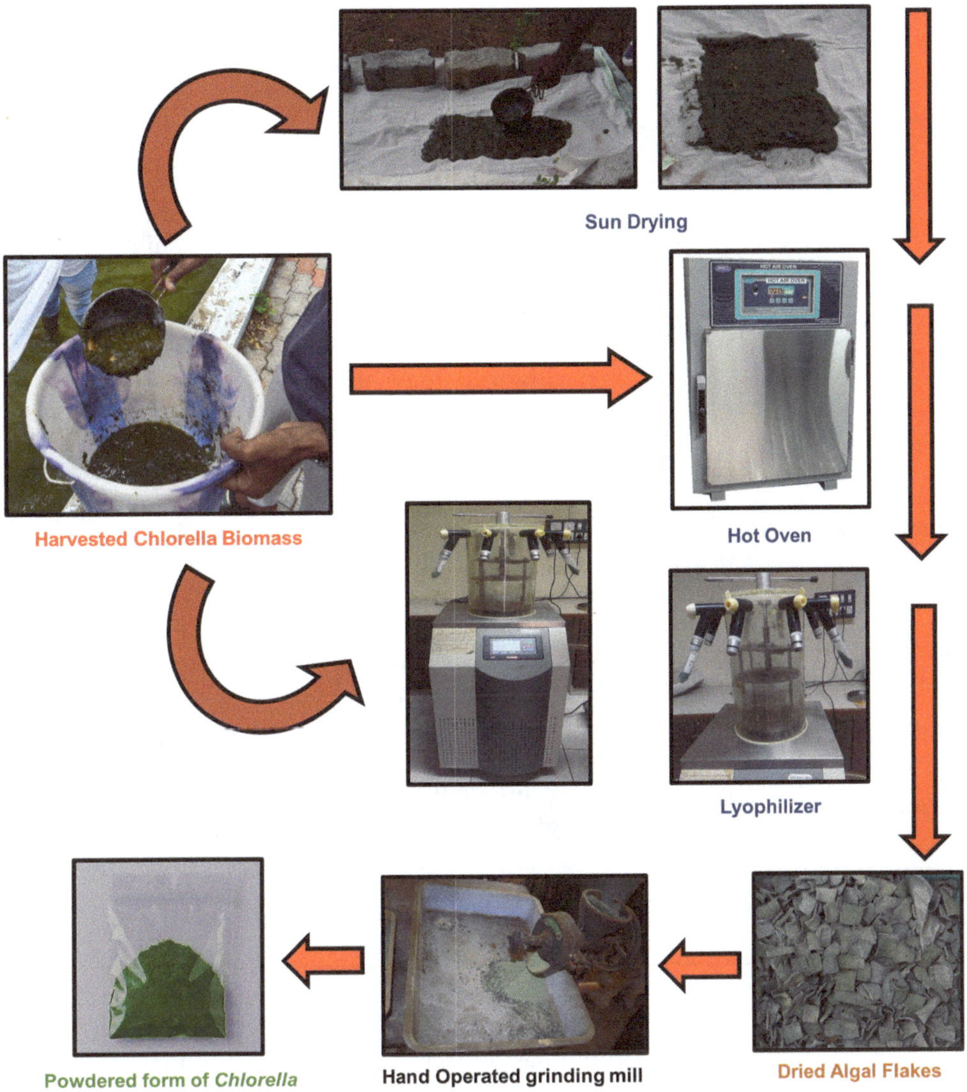

Fig. 1 Schematic illustration showing the conversion of wet algal culture into dried algal powder

3.1 Drying of Chlorella *Biomass*

3.1.1 Direct Sunlight

1. Place a clean cloth over a sand band and thinly apply the concentrated culture.
2. Cover it with mesh to allow it to dry in the sun and prevent bug infestation.

3.1.2 Hot Air Oven

1. Spread the concentrated biomass as a thin layer on a clean heat resistant tray.
2. Set the Temperature between 60 and 80 °C until constant weight is achieved.
3. Periodically check and turn the biomass for uniform drying.

3.1.3 Freezing Drying

1. Freeze the concentrate *Chlorella* biomass at −20 °C or below.
2. Transfer it to a Lyophilizer and dry under vacuum until complete moisture in removed.
3. Allow the dried biomass to form pellets or brittles flakes, then use a grinding mill to gently grind the dried *Chlorella* pellets in a fine powder.
4. Grind gently in short bursts to avoid heat buildup and sieve to desired particle size.
5. Record the dry weight and calculate the yield.

$$\text{Yield} = (\text{dry mass}/\text{initial wet mass}) \times 100\%$$

6. Store the grinded powdered in an airtight container with desiccant and label with date and dry weight.

3.2 Observation

The *Chlorella* biomass gradually loses moisture during the drying process. As a result, weight, volume, and wet texture all decrease significantly. The biomass appears sticky and dark green in the early stages. However, it becomes dry, brittle, and lighter as it dries (Fig. 1). Oven-dried samples often dry more evenly and quickly. Sun-dried biomass may exhibit color differences due to unequal exposure. Freeze-dried biomass frequently retains its porous character and bright green color, indicating improved pigment and heat-sensitive component preservation. When the material reaches a constant weight, the flakes feel dry to the touch and are easily broken. Powdering produces a fine, free-flowing green powder with a uniform particle size.

4 Notes

1. Spread biomass in thin layers for faster and even drying.
2. Stir or turn biomass periodically to improve uniform drying.
3. Ensure biomass is completely dry before grinding to avoid clumping.
4. Keep the powder away from light and humidity to maintain quality.
5. Use freeze-drying for high-value pigment or nutraceutical applications.
6. Record both wet and dry weights for accurate yield calculation.

References

1. Madhubalaji CK, Mudaliar SN, Chauhan VS, Sarada R (2021) Evaluation of drying methods on nutritional constituents and antioxidant activities of *Chlorella vulgaris*. J Appl Phycol 33(3):1419–1434
2. Fayaazuddin T, Prakash P, Shakena Fathima T, Dhanasekaran D (2023) Commercial astaxanthin production from green alga Haematococcus pluvialis. In: Food microbiology-based entrepreneurship: making money from microbes. Springer, Singapore, pp 279–304
3. Karaaslan S, Uysal O, Uysal FO, Ekinci K, Kumbul BS (2016) Mathematical modelling of drying of *Chlorella* sp. at different conditions. Eur J Sustain Dev 5(4):421–430
4. Yerima M et al (2018) Effects of sun and oven drying on quality of oil produced from *Chlorella vulgaris*. Biotechnol J Int 22(1):1–10
5. Chen X, Ma Q, Li L, Qin S (2013) Study on drying characteristics of microalgae under different conditions. Adv Mater Res 724–725:296–299

Part V

Value Added Products from *Chlorella*

Chapter 25

Fish Feed Formulation with *Chlorella* Wet Paste Form (Live/Concentrated Feed)

Mano Govindharaj, Kaliyan Barathikannan, Sai Kumar Tammina, Fawzi Banat, Srinivas Mettu, Manideep Pabba, and Ramachandran Chelliah

Abstract

This chapter outlines an experimental validation protocol for formulating fish feed using *Chlorella* wet paste (live or concentrated biomaterial) and assessing its effects on tilapia. The study investigates the impact of varying *Chlorella* wet paste inclusion levels on growth performance, feed conversion ratio (FCR), health biomarkers, gut morphology, antioxidant defenses, and immune response. A randomized complete design (CRD) with multiple inclusion levels ensures statistical reliability. Detailed methods for wet paste incorporation, pellet processing, biochemical analyses, histological measurements, and troubleshooting are provided to facilitate the accurate replication of the experiment.

Key words *Chlorella* wet paste, Live microalgae feed, Tilapia, Alternative protein, Aquafeed formulation

1 Introduction

In aquaculture, producers primarily rely on formulated feeds to ensure the health, optimal growth, and quality of farmed fish [1]. In aquaculture specifically, fishmeal and soybean meal have historically served as the primary protein sources due to their balanced amino acid profiles and high digestibility [2]. However, the rapid expansion of global aquaculture has intensified the demand for these ingredients, resulting in rising costs, environmental pressures associated with wild fish harvesting, and increasing competition with terrestrial livestock industries. These challenges have prompted researchers and feed manufacturers to explore sustainable, locally available, and nutritionally efficient alternatives to conventional protein sources. Microalgae have emerged as one of the most promising candidates in this context. Among them, *Chlorella* spp. Stand out due to their exceptional nutrient composition,

Dhanasekaran Dharumadurai (ed.), *Chlorella Genomics, Cultivation, and Bioenterprise: Green Economy through Algal Innovations*, Springer Protocols Handbooks, https://doi.org/10.1007/978-1-0716-5276-3_25,
© The Author(s), under exclusive license to Springer Science+Business Media, LLC, part of Springer Nature 2026

including 45–55% protein, polyunsaturated fatty acids, vitamins, minerals, pigments, and bioactive polysaccharides, making them a high-value ingredient for modern aquafeeds. [3] In addition to serving as a direct protein substitute, *Chlorella* contributes unique functional properties, including antioxidant enhancement, immune modulation, improved gut morphology, and stress resilience, in farmed fish.

Traditionally, *Chlorella* has been incorporated into feeds in its dried meal form. However, the drying process can degrade heat-sensitive bioactive compounds, reduce pigment stability, and increase production costs [4]. In contrast, this protocol provides a comprehensive step-by-step workflow for the formulation of *Chlorella* wet paste (live or concentrated microalgal biomass) that maintains cellular integrity and preserves functional nutrients, including chlorophylls, carotenoids, vitamins, and natural enzymes. Wet paste also offers advantages in palatability, freshness, and reduced processing energy, making it an attractive ingredient for hatchery and grow-out feeds. The formulations are fed to assess their effects on nutritional performance, physiological health, immune and antioxidant responses, and gastrointestinal morphology. By integrating wet paste microalgae into practical feed formulations, this study contributes to the development of environmentally sustainable and nutritionally enhanced aquafeeds aligned with circular bioeconomy principles.

The methodology is suitable for research laboratories, feed manufacturers, and academic institutions working in aquaculture nutrition. The protocol ensures reproducibility and supports the optimization of feed formulation for improved fish health and performance.

2 Materials

2.1 Fish Feed Formulation

1. *Protein sources.*
   (a) Fishmeal.
   (b) *Chlorella* biomass (meal or wet paste, depending on treatment).
2. *Carbohydrate and energy sources.*
   (a) Corn starch and/or wheat flour.
   (b) Rice bran.
3. *Lipid source.*
   (a) Fish oil or vegetable oil.
4. *Minerals and additives.*
   (a) Dicalcium phosphate.
   (b) Vitamin and mineral mixture.
   (c) Natural binder (e.g., wheat gluten, pea starch, and guar gum).

5. *Digestibility marker (for diets used in the digestibility trial).*
   (a) Chromic oxide (Cr_2O_3), analytical grade.
6. *Water.*
   (a) Clean freshwater for dough preparation.

2.2 Equipment

1. Precision balance (±0.01 g).
2. Mechanical mixer or food mixer.
3. Meat grinder or laboratory pellet mill with 2.5–3.0 mm die.
4. Trays for drying.
5. Drying oven with forced air circulation (capable of maintaining ≤50 °C).
6. Sieve (<500 μm) for grinding ingredients.
7. Vacuum sealer and opaque plastic bags.
8. Refrigerator (4 °C) for feed storage.

3 Method

3.1 Formulation

1. Decide target nutrient levels (e.g., 40% crude protein, 9% crude lipid, and a specific digestible protein/digestible energy ratio).
2. Choose test diets with graded *Chlorella* inclusion, for example:
   (a) Diet 1: 0% *Chlorella* (fishmeal-based reference).
   (b) Diet 2: 30% *Chlorella.*
   (c) Diet 3: 60% *Chlorella.*
   (d) Diet 4: 100% *Chlorella* as the main protein source.
3. Use apparent digestibility values (from the digestibility trial or literature, e.g., Lupatsch table) for fishmeal and *Chlorella* to calculate digestible protein and digestible energy and adjust inclusion levels so that all diets have comparable digestible nutrient profiles.
4. Prepare a formulation table according to Table 1.

3.2 Feed Preparation for Formulations

Step 1: Weighing ingredients

1. Weigh all dry ingredients (fishmeal, *Chlorella* meal or dried paste, starch/flour, dicalcium phosphate, vitamin–mineral premix, binder) separately according to the formulation (g/kg as fed).
2. If using *Chlorella* wet paste, determine its dry matter content and adjust the weighed amount and added water so that the final dough moisture is ~25–30%.

Table 1
Formulation and nutrient profile of experimental diets

Component/ parameter	Diet FM (Fishmeal)	Diet C30 (30% *Chlorella*)	Diet C60 (60% *Chlorella*)	Diet C100 (100% *Chlorella*)
Ingredient inclusion for the formulation of fish feed				
Fishmeal	640	420	210	–
Chlorella biomass	–	260	520	780
Corn starch	300	250	180	120
Vitamin–mineral premix	10	10	10	10
Dicalcium phosphate	–	10	30	50
Vegetable oil	20	20	20	20
Binder	30	30	30	20

Step 2: Grinding

1. Pass all solid ingredients through a < 500 μm sieve to achieve a uniform particle size.
2. Break any lumps in *Chlorella* biomass or fishmeal to ensure homogeneous mixing.

Step 3: Dry mixing

1. Transfer all dry ingredients (except oil) into a mechanical mixer.
2. Mix for 5–10 min until the blend appears uniform in color and texture.
3. If chromic oxide is included as an indigestible marker, premix it thoroughly with a small portion of ground ingredient first, then add to the main batch to avoid clumping.

Step 4: Adding liquids

1. Add the oil (fish or vegetable) slowly while the mixer is running to coat the particles evenly.
2. Gradually add water (and *Chlorella* wet paste if used) to the mixture while mixing.
3. Continue mixing until a cohesive dough form with a moisture content of approximately 25–30%, suitable for extrusion.

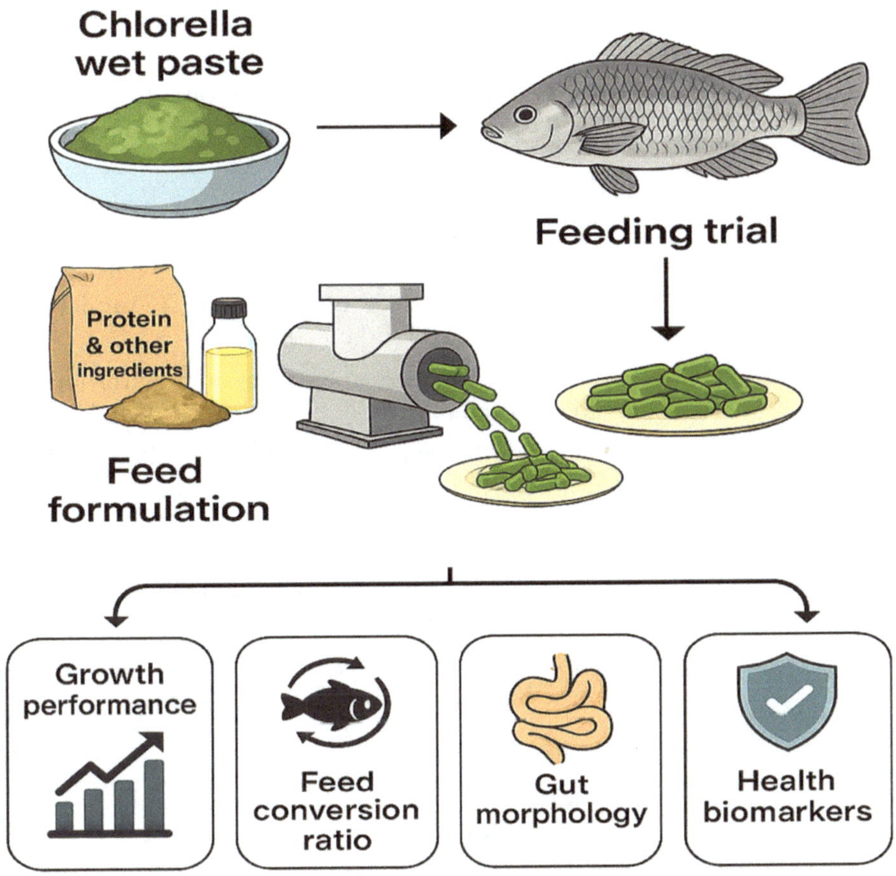

Fig. 1 Overall protocol for incorporating *Chlorella* wet paste (live or concentrated biomass) into formulated aquafeed for Nile tilapia. The workflow includes ingredient weighing, blending of wet paste with dry components, dough conditioning, extrusion/pellet formation, controlled drying, and vacuum packaging. The formulated diets are then subjected to biological validation through growth performance analysis, feed conversion ratio (FCR) assessment, biochemical profiling (antioxidant and immune biomarkers), and histological evaluation of gut morphology. The figure summarizes key experimental stages, highlighting how the use of fresh *Chlorella* biomass preserves bioactive compounds and enhances feed functionality

Step 5: Dough conditioning

1. Allow the dough to rest for 5–10 min to ensure binder hydration and uniform moisture distribution.
2. Check consistency: the dough should be firm enough to hold shape but soft enough to be extruded without cracking.

Step 6: Pelleting/extrusion

1. Fit the meat grinder or pellet mill with a 2.5–3.0 mm die.
2. Feed the conditioned dough into the grinder and extrude into continuous strands.

3. Cut strands into pellets of the desired length (typically 2–4 mm for juvenile tilapia).
4. At this stage, the pellets are moist, fragile, and handled with care.

Step 7: Drying

1. Spread pellets in a single layer on drying trays to avoid clumping.
2. Dry in a forced-air oven at ≤50 °C until the moisture content is <100 g/kg (10%).
3. Gently mix or turn the pellets halfway through drying to ensure even drying.
4. Confirm dryness by checking that pellets are firm and do not deform under light pressure (Fig. 1).

Step 8: Cooling and sieving

1. Remove trays from the oven and allow pellets to cool to room temperature.
2. Sieve to remove fines and broken pieces; these can be reused in the next batch if desired [5]..

Step 9: Storage

1. Pack cooled pellets into opaque, airtight bags (preferably vacuum-sealed).
2. Label each bag with diet code, batch number, and production date.
3. Store at 4 °C until use to minimize oxidation of lipids and degradation of pigments and vitamins.

4 Quality Check and Stability

1. *Water stability test:*
   (a) Place a known amount of pellets in water at the experimental temperature.
   (b) Observe physical integrity for up to 24 h, and the pellets from this formulation should remain stable over this period with minimal disintegration [6].
2. *Proximate analysis:*
   (a) Determine moisture, crude protein, crude lipid, ash, and gross energy to verify that the final composition matches the target formulation [5].

3. *Digestible nutrient calculation (optional):*
   (a) Combine proximate data with apparent digestibility coefficients from the digestibility trial (fishmeal vs. *Chlorella*) to calculate digestible protein and digestible energy for each diet [7].

5 Notes

1. If pellets are too soft, disintegrate quickly: Increase binder level or reduce added water, and check for drying temperature and time.
2. If the pellets are too hard and brittle, slightly increase the water content during dough formation or reduce drying time.
3. Uneven green color in *Chlorella* diets, improve dry mixing, and ensure *Chlorella* is finely ground and well dispersed.
4. If odor comes during storage, reduce storage time, use fresh oil, and consider adding an antioxidant (e.g., vitamin E).
5. If the formulated pellets are large and vary in size, maintain a constant feed rate into the grinder and cut pellets uniformly during extrusion.

References

1. Prabu E, Felix S, Felix N, Ahilan B, Ruby P (2017) An overview on significance of fish nutrition in aquaculture industry. Int J Fish Aquat Stud 5:349–355
2. Gasco L, Gai F, Maricchiolo G, Genovese L, Ragonese S, Bottari T et al (2018) Fishmeal alternative protein sources for aquaculture feeds. In: Feeds for the aquaculture sector: current situation and alternative sources. Springer, pp 1–28
3. Ampofo J, Abbey L (2022) Microalgae: bioactive composition, health benefits, safety and prospects as potential high-value ingredients for the functional food industry. Foods 11:1744
4. Madhubalaji CK, Mudaliar SN, Chauhan VS, Sarada R (2021) Evaluation of drying methods on nutritional constituents and antioxidant activities of Chlorella vulgaris cultivated in an outdoor open raceway pond. J Appl Phycol 33:1419–1434
5. Thangaraj P (2015) Proximate composition analysis. In: Pharmacological assays of plant-based natural products. Springer, pp 21–31
6. Lim C, Cuzon G (1994) Water stability of shrimp pellet: a review. Asian Fish Sci 7:115–126
7. Barone RSC, Sonoda DY, Lorenz EK, Cyrino JEP (2018) Digestibility and pricing of Chlorella sorokiniana meal for use in tilapia feeds. Sci Agric 75:184–190

Chapter 26

Production of *Chlorella* Dry Powder as a Value-Added Supplemental Feed

Kaliyan Barathikannan, Mano Govindharaj, Fawzi Banat, Srinivas Mettu, and Ramachandran Chelliah

Abstract

Chlorella biomass represents a sustainable, nutrient-rich feed ingredient with significant potential in aquaculture, poultry, and livestock nutrition. This chapter outlines a practical, scalable protocol for producing high-quality *Chlorella* dry powder, beginning with biomass harvesting and followed by washing, dewatering, drying, milling, and packaging. The process emphasizes maintaining nutritional integrity, particularly proteins, lipids, pigments, and antioxidants, while ensuring product stability and flowability. Different drying methods (freeze-drying, tray drying, spray-drying) are compared in terms of moisture content, nutrient retention, and suitability for laboratory or industrial applications. Key quality parameters such as moisture, color, odor, and microbial load are identified to ensure product safety and consistency. Troubleshooting guidelines and performance expectations provide a comprehensive framework for researchers and industry practitioners seeking to generate value-added algal feed products for diversified food-security sectors.

Key words *Chlorella* biomass, Value-added feed, Spray-drying, Freeze-drying, Functional feed ingredient, Protein-rich microalgae, Livestock supplementation, Biomass processing

1 Introduction

Microalgae have emerged as a transformative resource in sustainable food systems due to their high productivity, minimal land requirements, and rich nutritional profile. Among the various microalgal species, *Chlorella* is one of the most widely cultivated owing to its fast growth rate, robust cell structure, and substantial levels of protein (40–60%), essential fatty acids, vitamins, minerals, pigments, and bioactive compounds. These attributes make *Chlorella* an attractive ingredient for formulating supplemental feed in aquaculture, poultry, and small-scale livestock sectors. As global demand rises for alternative protein sources and environmentally efficient feed solutions, microalgae offer a strategic advantage.

Dhanasekaran Dharumadurai (ed.), *Chlorella Genomics, Cultivation, and Bioenterprise: Green Economy through Algal Innovations*, Springer Protocols Handbooks, https://doi.org/10.1007/978-1-0716-5276-3_26,
© The Author(s), under exclusive license to Springer Science+Business Media, LLC, part of Springer Nature 2026

Chlorella can be cultivated in controlled photobioreactors or open ponds using CO_2-rich aeration, wastewater, or nutrient-recycled media, aligning well with circular bioeconomy and carbon-reduction targets. However, the successful incorporation of micro-algae into feed systems depends heavily on the production of a stable, high-quality dry powder with consistent nutritional characteristics and longer shelf life. Drying is one of the most critical steps in converting wet biomass (typically 90–95% moisture) into a storable, flowable powder. The method selected—freeze-drying, tray/cabinet drying, or spray-drying—influences pigment preservation, protein quality, lipid oxidation, and overall sensory properties. Therefore, optimizing the drying conditions is essential to maintain the functional properties of *Chlorella*, including its antioxidant capacity, chlorophyll content, and digestibility. This chapter presents a detailed, stepwise protocol for producing *Chlorella* dry powder, beginning from biomass harvesting through final quality evaluation. Each staged watering, drying, milling, sieving, and packaging—is described along with practical operational parameters suitable for research laboratories and pilot-scale facilities. Quality criteria such as moisture content, color stability, microbial safety, and nutrient integrity are outlined to ensure that the final powder meets requirements for use as a value-added supplemental feed. Moreover, troubleshooting guidance is provided to address common challenges such as clumping, browning, low yield, and lipid oxidation. By enabling reproducible, scalable production, this protocol supports the effective utilization of *Chlorella* biomass in modern feed and food-security frameworks.

2 Materials

1. *Chlorella* culture (late exponential/early stationary phase).
2. Freshwater or salt-adjusted water (strain-dependent).
3. Optional: food-grade anticaking agent (0.5–1%), antioxidant such as vitamin E (0.1–0.5%).
4. Cleaning agents (70% ethanol, mild detergent).
5. Moisture-barrier pouches or airtight bottles.
6. Labels (batch code, date, moisture content).

Equipment

7. Culture tanks or photobioreactor.
8. Centrifuge/membrane filtration unit/DAF system.
9. Filter press or cloth filters.
10. Drying system (freeze-dryer, tray/cabinet dryer, or spray dryer).

11. Grinder or hammer mill.
12. Sieves (250–500 μm).
13. Moisture analyzer or drying oven.
14. Analytical balance.
15. Thermometer and hygrometer.

3 Methods

Harvesting

1. Harvest culture at high cell density ($OD_{680} \approx 0.8–1.5$).
2. Stop aeration briefly to reduce foam.
3. Collect biomass:
   - *Centrifugation*: 4000–6000× *g*, 10–15 min at ≤25 °C.
   - *Alternative*: membrane filtration or DAF until slurry reaches ~20–50 g/L solids.
   - Decant supernatant and weigh wet biomass.

Washing and Dewatering

1. Resuspend pellet in equal volume of clean water to remove media components.
2. Centrifuge again (same parameters).
3. Dewater using centrifuge or filter press to obtain thick paste (~10–25% solids).
4. Measure moisture using oven-drying or a moisture analyzer.

Drying the Biomass (Target Moisture: ≤8–10%)

Option A: Freeze-Drying

1. Spread paste in 3–5 mm layers on trays and freeze at −20 to –40 °C.
2. Lyophilize under vacuum (0.1–0.5 mbar) for 24–48 h.
3. Confirm moisture ≤10%.

Option B: Tray/Cabinet Drying

1. Spread paste in thin layers (2–4 mm).
2. Dry at 40–60 °C with adequate airflow for 6–12 h.
3. Stop when biomass is brittle; confirm moisture ≤10%.

Option C: Spray-Drying

1. Dilute paste to 10–15% solids for pumpability.
2. Typical conditions: inlet 150–180 °C, outlet 70–90 °C.
3. Collect powder from cyclone; confirm moisture 4–8%.

Milling and Sieving

- Break dried biomass into smaller fragments.
- Mill using grinder/hammer mill; avoid overheating.
- Sieve to 250–500 μm for uniform particle size.
- Regrind any oversize fraction.

Optional Additive Incorporation

- Add anticaking agent (0.5–1%) or antioxidant (0.1–0.5%) and mix thoroughly until uniform.

Packaging and Storage

- Fill powder into airtight, moisture-barrier containers.
- Label with batch number, date, and moisture content.
- Store in cool, dry, dark conditions (≤25 °C, RH <60%).
- Avoid repeated opening of bulk containers.

Quality Checks

- Moisture ≤8–10%.
- Color: Green to deep green, free of clumps.
- Odor: Clean algal smell (no sour/rancid notes).
- Optional: Microbial quality (TPC, yeast/mold, Salmonella-free).
- Optional nutritional analysis: Protein, lipids, ash, chlorophyll.

4 Notes

Implementation of the *Chlorella* drying protocol produced a stable, high-quality dry biomass with consistent physicochemical characteristics appropriate for supplemental feed applications [1, 2]. Freeze-drying and spray-drying resulted in the most uniform and fine powders, whereas tray/cabinet drying delivered slightly coarser particles with moderate pigment darkening [2, 4].

Key observations include the following:

- *Moisture content* of final powders ranged between *4% and 10%*, depending on the drying method [2, 5].

- *Freeze-dried biomass* showed excellent color retention (bright green), minimal oxidation, and the highest preservation of chlorophyll, carotenoids, and protein integrity [1, 6].
- *Spray-dried biomass* exhibited uniform particle size (100–300 μm) and good flowability, suitable for pilot and industrial scales [4, 10].
- *Tray-dried samples* showed moderate color darkening and slightly higher oxidation due to heat exposure but maintained acceptable nutritional quality [2, 10].
- *Milling and sieving* produced particle sizes between *250 and 500 μm*, ideal for feed mixing [8, 11].
- *Optional antioxidant incorporation* reduced pigment oxidation during storage [6].
- *Microbial quality* (TPC, yeast/mold) remained within acceptable limits when drying temperature and storage humidity were controlled [1, 2].

Overall, all three drying approaches produced functional, shelf-stable powders when the protocol's moisture and handling conditions were followed [1, 2, 4].

Inferences

- *Drying temperature strongly influences nutrient retention.*

  Freeze-drying preserves pigments and heat-labile compounds best, whereas higher temperatures in tray drying moderately reduce color and slight nutrient quality [1, 2, 10].
- *Moisture content is the key determinant of flowability and shelf stability.*

  Powders with ≤8% moisture exhibit minimal clumping, reduced microbial risk, and extended storage stability [2, 5].
- *Spray-drying offers the most scalable and industry-suitable method.*

  It balances moisture control, color retention, uniform particle size, and operational efficiency [4, 10].
- *Milling and sieving standardize texture and enhance feed blending compatibility.*

  Overheated grinding reduces PUFA stability; therefore, low-heat milling is essential [3, 8].
- *Use of antioxidants* (e.g., *vitamin E) improves oxidative stability during storage.*

  Antioxidants prevent rancidity and pigment degradation, especially in lipid-rich *Chlorella* strains [6, 7].

- *Proper packaging and storage conditions are critical.*

  Moisture-barrier pouches and controlled temperature (<25 °C) minimize oxidation, moisture gain, and nutrient losses [1, 2].
- *High cell-density harvesting increases final yield and reduces processing costs.*

  Biomass collected at $OD_{680} > 0.8$ consistently yields higher solids during dewatering [4, 5, 12].
- These inferences confirm that optimized drying and handling parameters directly enhance product quality and functional utility as a supplemental feed ingredient [1, 2, 6].

Observation

When this protocol is performed correctly, the following outcomes should be expected:

1. *Physical Quality.*
   - Fine, uniform green powder with no visible clumps or moisture spots.
   - Particle size between *250 and 500 μm.*
   - Good flowability (no sticking or agglomeration).
2. *Chemical Quality.*
   - Moisture content ≤*8–10%.*
   - Minimal lipid oxidation (no rancid odor).
   - High pigment retention (chlorophyll and carotenoids) depending on drying method.
   - Preservation of protein and PUFA content.
3. *Microbial Quality.*
   - Acceptable TPC (low counts for dried biomass).
   - Yeast/mold-free.
   - No pathogenic contamination (Salmonella-free).
4. *Sensory Indicators.*
   - *Color*: bright to deep green.
   - *Odor*: clean, fresh algal smell.
   - *Texture*: fine, free-flowing powder.
5. *Storage Stability.*
   - Stable for *3–6 months* when stored at <25 °C and <60% RH.
   - No clumping, browning, or off-odors during storage.
   - Quality maintained inside sealed moisture-barrier packaging.

Table 1
Summary of expected results under different drying methods

Drying method	Final moisture (%)	Color retention	Nutrient retention	Best use scale
Freeze-drying	4–8	Excellent	Very high	Lab/small scale
Tray/cabinet drying	6–10	Moderate	Good	Lab/medium scale
Spray-drying	4–8	Good	Good–high	Pilot/industrial

6. *Process Efficiency.*
   - Yield increases with high-density harvesting (20–50 g/L solids).
   - Freeze-drying: Highest retention, highest cost.
   - Spray-drying: High retention, industry scalable.
   - Tray drying: Cost-effective, moderate retention.

These expected results help define quality benchmarks for value-added *Chlorella* dry powder (Table 1).

Troubleshooting (Five Practical Tips)

1. Clumping/poor flowability: Caused by high moisture or humid storage; fix by increasing drying time, using moisture-barrier packaging, or adding 0.5–1% anticaking agent.
2. Powder turns brown: Caused by overheating during drying; fix by keeping tray drying ≤60 °C and minimizing exposure to light/oxygen.
3. Low final yield: Caused by low culture density or biomass losses; fix by harvesting at peak density, increasing centrifugation efficiency, and recovering biomass from equipment surfaces.
4. Off-odors (sour/rancid): Caused by lipid oxidation or microbial growth in wet biomass; fix by drying immediately after harvest, ensuring final moisture <10%, and adding antioxidants.
5. Spray dryer nozzle clogging: Caused by overly viscous slurry or large particulates; fix by diluting feed to 10–15% solids, pre-filtering the slurry, and cleaning the nozzle regularly (Fig. 1).

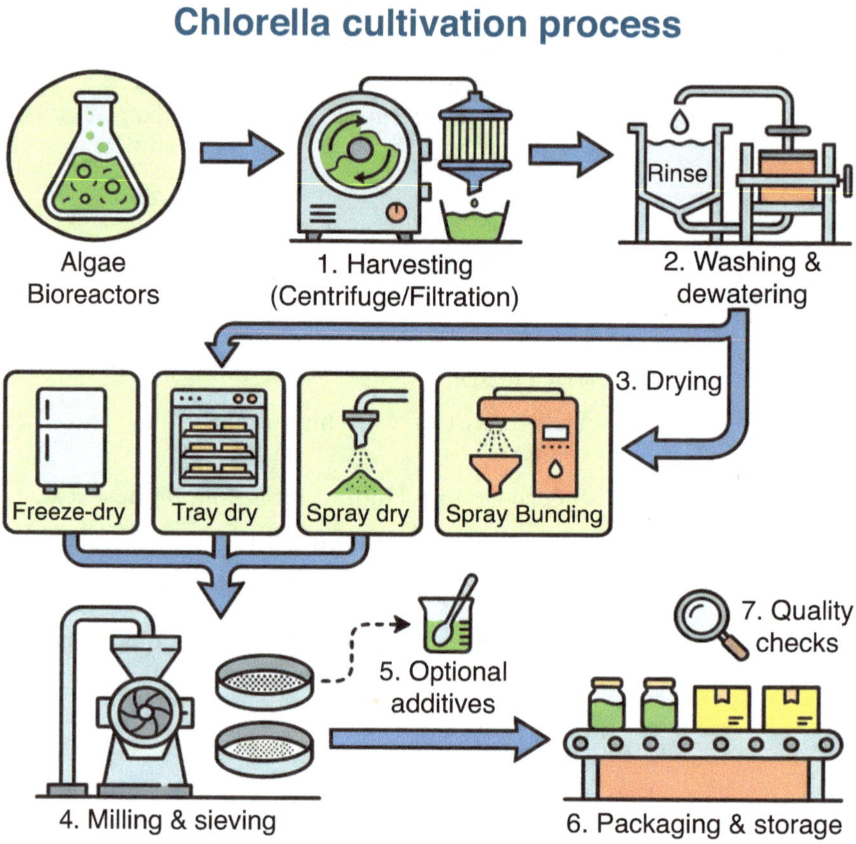

Fig. 1 *Chlorella* cultivation and processing workflow for dry powder production: The processing pipeline used to convert *Chlorella* biomass into stable dry powder suitable for supplemental feed applications. The workflow begins with biomass generation in algae bioreactors, followed by (1) harvesting through centrifugation or filtration, and (2) washing and dewatering to remove media residues and concentrate the slurry. The biomass is then subjected to (3) drying using one of several methods—freeze-drying, tray/cabinet drying, spray-drying, or spray bundling—depending on scale and nutrient-retention requirements. Dried biomass is subsequently (4) milled and sieved to achieve uniform particle size, after which (5) optional additives such as anticaking agents or antioxidants may be incorporated. The final product undergoes (6) packaging and storage in moisture-barrier containers, followed by (7) quality checks including moisture analysis, sensory inspection, and microbial assessment. This process ensures consistent, high-quality *Chlorella* powder for aquaculture, poultry, and livestock feed formulations

References

1. Becker EW (2013) Microalgae for human and animal nutrition. In: Handbook of microalgal culture: applied phycology and biotechnology, pp 461–503. https://doi.org/10.1002/9781118567166
2. Safi C, Zebib B, Merah O, Pontalier PY, Vaca-Garcia C (2014) Morphology, composition, production, processing and applications of Chlorella vulgaris: a review. Renew Sust Energ Rev 35:265–278
3. Ryckebosch E, Muylaert K, Foubert I (2012) Optimization of an analytical procedure for extraction of lipids from microalgae. J Am Oil Chem Soc 89(2):189–198
4. Liu J, Hu Q (2013) Chlorella: industrial production of cell mass and chemicals. In: Handbook of microalgal culture: applied phycology

and biotechnology, pp 327–338. https://doi.org/10.1002/9781118567166.ch16

5. Pragya N, Pandey KK, Sahoo PK (2013) A review on harvesting, oil extraction and biofuels production technologies from microalgae. Renew Sust Energ Rev 24:159–171
6. Buono S, Langellotti AL, Martello A, Rinna F, Fogliano V (2014) Functional ingredients from microalgae. Food Funct 5(8):1669–1685
7. Halim R, Gladman B, Danquah MK, Webley PA (2011) Oil extraction from microalgae for biodiesel production. Bioresour Technol 102(1):178–185
8. Bleakley S, Hayes M (2017) Algal proteins: extraction, application, and challenges concerning production. Foods 6(5):33
9. Ferjančič B, Skrt M, Korošec M, Bertoncelj J (2022) Comparative analysis of dietary fibre determination by AOAC 991.43 and AOAC 2011.25 for frequently consumed foods in Slovenia. Food Chem 397:133753
10. da Silva Sousa VE (2020) Evaluation of ohmic heating processing on extraction of bioactive compounds from microalgae biomass. Master's thesis, Universidade do Minho, Portugal
11. Alhattab M, Kermanshahi-Pour A, Brooks MSL (2019) Microalgae disruption techniques for product recovery: influence of cell wall composition. J Appl Phycol 31(1):61–88
12. Gautam R, Patial SK, Singh S (2024) Algae biomass: importance, harvesting techniques, extraction methods, and associated challenges. In: Value added products from bioalgae based biorefineries: opportunities and challenges. Springer Nature Singapore, Singapore, pp 67–94

Chapter 27

Chlorella-Enriched Feed Pellet Preparation

Ramachandran Chelliah and Deog-Hwan Oh

Abstract

Enriched pellet preparation involves the formulation and processing of nutritionally balanced feed supplemented with vitamins, *Chlorella* minerals, probiotics, and bioactive extracts. The present protocol details the preparation of laboratory-scale feed pellets suitable for both terrestrial and aquatic organisms. It describes formulation, mixing, palletization, drying, and storage steps, ensuring uniform distribution of enrichment additives and retention of bioactivity. The chapter also includes troubleshooting and error identification notes for maintaining reproducibility and feed quality.

Key words Feed formulation, Pelletization, Probiotics, *Chlorella*, Enrichment, Binder, Drying, Stability, Troubleshooting

1 Introduction

Feed enrichment technology provides a platform for delivering targeted nutrients and bioactives efficiently to animals. Unlike conventional mash feeds, pelletized feeds offer better nutrient stability, lower wastage, and improved digestibility. Incorporating enrichment agents such as vitamins, probiotics, or plant extracts helps improve gut health, feed conversion ratio (FCR), and immunity. This chapter describes a standardized bench-scale procedure to prepare enriched pellets from commonly available feed ingredients. Emphasis is placed on uniform blending, optimal moisture control, temperature management during drying, and preservation of heat-sensitive additives like probiotics.

2 Materials

Prepare all reagents using analytical-grade chemicals and ensure all feed ingredients are free of rancidity, mold, or insect infestation. Use clean, food-grade containers and tools throughout the process.

Dhanasekaran Dharumadurai (ed.), *Chlorella Genomics, Cultivation, and Bioenterprise: Green Economy through Algal Innovations*, Springer Protocols Handbooks, https://doi.org/10.1007/978-1-0716-5276-3_27,
© The Author(s), under exclusive license to Springer Science+Business Media, LLC, part of Springer Nature 2026

2.1 Feed Ingredients

1. Protein source: Fish meal, defatted soybean meal, insect meal, or groundnut cake.
2. Carbohydrate source: Wheat flour, maize flour, or rice bran.
3. Lipid source: Fish oil, soybean oil, or algal oil.
4. Binder: Sodium alginate, gelatin, or carboxymethyl cellulose (0.5–2% w/w).
5. Vitamin and mineral premix: Commercial mix (species-specific).
6. Probiotics or prebiotics: *Chlorella, Lactobacillus plantarum, Bacillus subtilis*, inulin, Fructooligosacharide (FOS), or Chit-ooligosacharide (COS), Inulin, Xylooligosacharide (XOS).
7. Functional additives: Seaweed hydrolysate, herbal extract (Moringa oleifera, Codonopsis lanceolata), or postbiotic biomass.
8. Moisture medium: Distilled water or water–oil mixture.

2.2 Equipment

1. Digital balance (accuracy ±0.01 g).
2. Planetary or mechanical mixer.
3. Pelletizer/extruder (2–4 mm die).
4. Oven or hot air dryer (40–60 °C).
5. Moisture analyzer.
6. Thermometer and hygrometer.
7. Sterile spatulas and trays.
8. Vacuum sealer and airtight pouches.

3 Methods

All procedures should be carried out at room temperature unless otherwise specified.

3.1 Feed Formulation

1. Calculate desired nutritional composition (e.g., 35% protein, 5–10% lipid, 6–8% moisture).
2. Weigh dry ingredients (e.g., fish meal 30%, soybean meal 25%, rice bran 25%, wheat flour 20%).
3. Mix ingredients for 10–15 min until homogeneous.
4. Sieve (<1 mm) to improve uniformity.

3.2 Enrichment Incorporation

1. Prepare enrichment suspension by dissolving premix, prebiotics, and extracts in 100–200 mL sterile water.
2. For probiotics, add *Chlorella* culture (<40 °C) to prevent thermal loss.

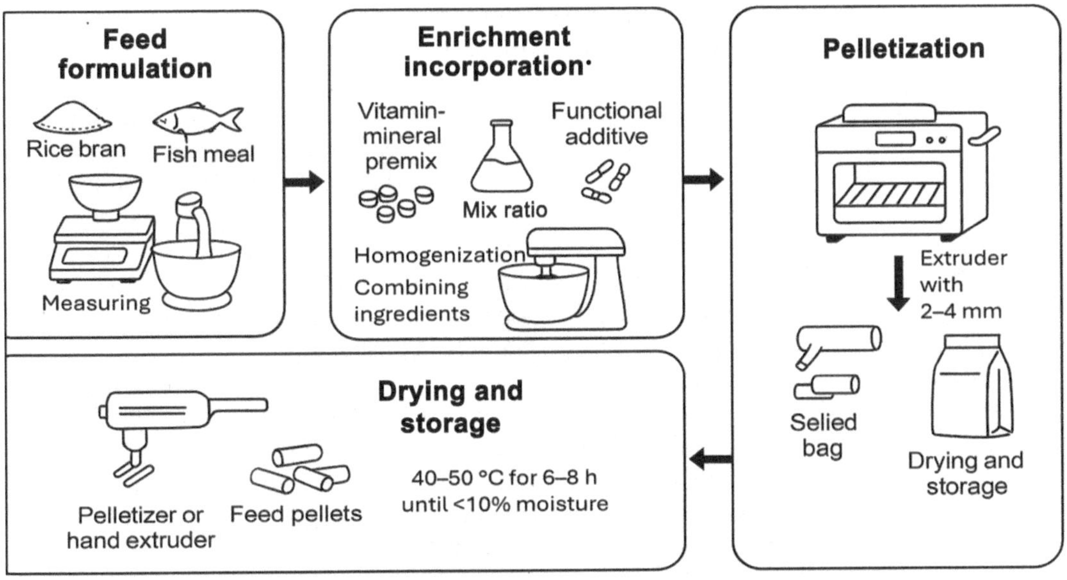

Fig. 1 Model on enriched feeding pellet preparation

3. Gradually add to dry mix while stirring until moisture reaches 25–30%.
4. Check dough consistency, smooth, nonsticky texture is ideal (Fig. 1).

3.3 Pelletization

1. Feed dough into pelletizer or extruder with 2–4 mm die.
2. Cut to 5–10-mm length and spread evenly on trays.
3. Ensure continuous extrusion and adjust moisture if flow is intermittent.

3.4 Drying

1. Dry at 40–50 °C for 6–8 h until <10% moisture.
2. Keep below 45 °C for probiotic feeds to maintain viability.
3. Stir every 2 h for uniform drying.
4. Cool to room temperature before packaging.

3.5 Storage

1. Vacuum-seal cooled pellets in polyethylene bags.
2. Store at 4 °C (≤1 month) or −20 °C (≤6 months).
3. Label with batch number, enrichment type, and date.

3.6 Quality Evaluation

1. Determine moisture content (105 °C for 3 h).
2. Analyze nutrient uniformity and texture.
3. Test probiotic viability via plate count (MRS agar).
4. Evaluate sensory quality and stability during storage.

The standardized protocol for enriched pellet preparation resulted in uniform, nutritionally stable feed pellets with consistent texture, density, and additive distribution. Moisture content after drying reliably decreased to <10%, ensuring shelf stability and inhibiting microbial growth. Pellets maintained structural integrity with minimal cracking when dried at 40–50 °C. Probiotic-enriched pellets retained 70–90% cell viability, depending on drying temperature and strain sensitivity. Functional enrichment additives such as herbal extracts, prebiotics, or postbiotics were evenly incorporated without visible phase separation. Pellets exhibited good mechanical strength and water stability, particularly when binders such as sodium alginate or carboxymethyl cellulose were used at 0.5–2%. Sensory evaluation confirmed acceptable aroma (nonrancid), color, and firmness during 1–2 months of refrigerated storage. Nutrient uniformity assays showed low batch-to-batch variation. Vacuum-packaged pellets displayed excellent stability with no mold growth or oxidation over short-term storage.

3.7 Inference

1. Binder concentration and moisture control are critical for achieving firm, durable pellets suitable for storage and handling.
2. Low-temperature drying (<45 °C) preserves the viability of heat-sensitive probiotics while maintaining pellet structure.
3. Uniform mixing prior to pelletization improves additive distribution, reducing variability in nutrients and bioactive delivery to animals.
4. Excess moisture leads to microbial growth and soft texture, emphasizing the importance of maintaining a drying endpoint of <10% moisture.
5. Vacuum packaging significantly prolongs shelf-life by reducing oxidative degradation and preventing moisture ingress.
6. Functional additives such as prebiotics, seaweed extracts, or herbal biomass remain stable when incorporated at the dough stage and protected from high temperatures.
7. Feed formulation flexibility allows tailoring pellets for species-specific nutritional needs, supporting enriched feeding strategies across aquaculture and terrestrial animal systems.

3.8 Expected Results

When the protocol is followed correctly, the following standard outcomes should be observed:

3.8.1 Physical Pellet Quality

1. Pellet diameter: 2–4 mm (depending on die).
2. Pellet length: 5–10 mm.
3. Surface: smooth, noncracked.

4. Texture: firm but not brittle.
5. Water stability: Pellets retain shape for >30–60 min (for aquatic feeds).

3.8.2 Chemical and Nutritional Properties

1. Final moisture content: <10%.
2. Uniform nutrient distribution with low variability across samples.
3. Stable lipid content with no rancidity (no off odors).
4. Minimal loss of heat-sensitive vitamins when temperature kept <50 °C.

3.8.3 Probiotic/Prebiotic Enrichment

1. Viability retention:
   (a) 70–90% for *Lactobacillus* spp.
   (b) 60–80% for *Bacillus* spp.
2. Prebiotics remain chemically stable throughout drying (Table 1).

3.8.4 Shelf Stability

1. No mold growth under refrigerated storage (4 °C) for 1 month or frozen storage (−20 °C) for up to 6 months.
2. No clumping or oil separation in vacuum-sealed bags.
3. Color remains consistent (no browning).

3.8.5 Sensory and Mechanical Evaluation

1. Acceptable aroma (fresh, nonrancid).
2. Hardness appropriate for intended species.
3. No powdery residue or pellet fragmentation during handling (*see* **Notes 1–6**).

Table 1
Notes and troubleshooting

S.No	Issue	Possible cause	Remedy
1	Pellets too soft	Low binder ratio	Increase binder to 0.5–1%
2	Cracking on drying	Rapid heating	Reduce temperature gradient
3	Sticky dough	Excess moisture	Add small amount of flour
4	Mold during storage	High residual moisture	Re-dry to <10% moisture
5	Loss of probiotic viability	High drying temperature	Dry below 45 °C or use encapsulation

4 Notes

1. Maintain aseptic conditions during probiotic addition [2, 3].
2. If essential oils or extracts are used, premix oil carriers before incorporation to prevent volatilization [3, 5].
3. Excess moisture (>35%) can lead to poor pellet texture and microbial spoilage [1, 5].
4. For aquatic feeds, binders like sodium alginate or gelatin improve water stability [1, 2].
5. For thermostable probiotics, encapsulation (e.g., alginate beads) may be used before mixing [3].
6. Record environmental parameters (temperature, humidity) during drying for reproducibility [4, 5].

References

1. National Research Council (NRC) (1993) Nutrient requirements of fish. National Academy Press, Washington, DC
2. Tsairidou S, Houston RD, Karapanagiotidis IT (2024) The importance of nutrition and selective breeding in aquaculture production. In: Aquatic food security. GB: CABI, pp 14–49
3. Madhulika, Meinam M, Deepti M, Ngasotter S, Gupta SS, Varghese T (2025) Functional feed additives in aquaculture to improve food security. In: Food security, nutrition and sustainability through aquaculture technologies. Springer Nature Switzerland, Cham, pp 375–396
4. Nguyen N, Jacobs M, Li J, Huang C, Li D, Navarro DM et al (2019) Technical note: concentrations of soluble, insoluble, and total dietary fiber in feed ingredients determined using Method AOAC 991.43 are not different from values determined using Method AOAC 2011.43 with the AnkomTDF Dietary Fiber Analyzer. J Anim Sci 97(9):3972–3983
5. Tacon AG, Metian M (2015) Feed matters: satisfying the feed demand of aquaculture. Rev Fish Sci Aquac 23(1):1–10

Chapter 28

Nutritional Analysis of *Chlorella* Aquafeed

Ramachandran Chelliah, Manideep Pabba, and Srinivas Mettu

Abstract

This protocol describes standardized procedures for evaluating the nutritional profile of *Chlorella*-enriched aquafeeds, including proximate composition, pigment content, amino acid profiling, mineral analysis, and antioxidant potential. The method is optimized for aquaculture feed formulations containing microalgal biomass and is suitable for feed quality assessment, formulation optimization, and research applications. This protocol provides complete procedures, quality control checkpoints, and troubleshooting steps to ensure accurate and reproducible results.

Key words *Chlorella*, Aquafeed, Proximate composition, Pigments, Amino acids, Minerals, Antioxidant activity, Nutritional analysis, Microalgae, Fish feed

1 Introduction

Chlorella spp. have gained significant attention in aquaculture nutrition due to their high protein content, balanced amino acid profile, polyunsaturated fatty acids, vitamins, minerals, pigments, and antioxidant compounds. Incorporating *Chlorella* into aquafeed formulations can enhance fish growth, feed conversion efficiency, immune modulation, gut health, and oxidative stress resilience. However, to ensure optimal feed quality, it is essential to perform standardized nutritional analysis of *Chlorella*-based aquafeeds. This protocol provides a comprehensive step-by-step workflow for analyzing the proximate composition, pigment content, amino acid profile, mineral composition, and antioxidant capacity of *Chlorella*-enriched aquafeed. The methodology is suitable for research laboratories, feed manufacturers, and academic institutions working in aquaculture nutrition, microalgae biotechnology, and fish physiology. The protocol ensures reproducibility and supports feed formulation optimization for improved fish health and performance.

Dhanasekaran Dharumadurai (ed.), *Chlorella Genomics, Cultivation, and Bioenterprise: Green Economy through Algal Innovations*, Springer Protocols Handbooks, https://doi.org/10.1007/978-1-0716-5276-3_28,
© The Author(s), under exclusive license to Springer Science+Business Media, LLC, part of Springer Nature 2026

2 Materials

2.1 Chemicals and Reagents

1. *Chlorella* powder (dried, <10% moisture).
2. Hexane (analytical grade).
3. Petroleum ether.
4. Sulfuric acid, sodium hydroxide, boric acid (for Kjeldahl).
5. Chloroform–methanol (2:1).
6. Folin–Ciocalteu reagent.
7. DPPH or ABTS reagent.
8. *Standards*:
   (a) BSA (protein standard).
   (b) Gallic acid (phenolics).
   (c) Trolox (antioxidant standard).
   (d) Amino acid standards (HPLC grade).
9. Nitric acid (HNO_3) for digestion.
10. Deionized water (DW).

2.2 Equipment

1. Analytical balance.
2. Soxhlet apparatus.
3. Kjeldahl digestion and distillation unit.
4. Muffle furnace.
5. UV–Vis spectrophotometer.
6. High-performance liquid chromatography (HPLC).
7. Atomic absorption spectrophotometer (AAS) or ICP-OES.
8. Centrifuge.
9. Mortar grinder or sample mill.
10. Drying oven (≤50 °C).
11. Water bath.

3 Methods

Below is the cleaned and connected method section including sample preparation, proximate analysis, pigments, amino acids, minerals, and antioxidant tests.

3.1 Sample Preparation

1. Grind *Chlorella* aquafeed pellets to <500 μm powder.
2. Dry samples at 105 °C to constant weight.
3. Store in airtight containers until analysis.

3.2 Proximate Composition Analysis

3.2.1 Moisture

1. Dry 5 g *Chlorella* sample at 105 °C, calculate moisture loss.

3.2.2 Crude Protein

1. Perform Kjeldahl digestion, distillation, and titration. Convert nitrogen to protein (N × 6.25).

3.2.3 Crude Lipid

1. Extract 2–3 g *Chlorella* sample using hexane in Soxhlet for 6 h.

3.2.4 Crude Fiber

1. Boil in diluted acid, then dilute alkali; dry, ash, and calculate fiber (Table 1).

3.2.5 Ash Content

1. Incinerate at 550 °C for 6 h.

3.2.6 Gross Energy

1. Measure using bomb calorimetry or compute using Atwater values.

3.3 Pigment Analysis (Chlorophyll and Carotenoids)

1. Extract *Chlorella* pigments using 80% acetone.
2. Measure absorbance at 663, 645, and 470 nm.
3. Calculate chlorophyll a, b, and carotenoids using Lichtenthaler equations (Table 2).

3.4 Amino Acid Profiling (HPLC)

1. Hydrolyze with 6 N HCl at 110 °C for 24 h.
2. Neutralize and filter.
3. Derivatize with OPA or PITC.

Table 1
Proximate composition—expected ranges and interpretation

Parameter	Expected result (typical range)	Scientific inference/outcome
Moisture	6–10%	Indicates proper drying; prevents microbial spoilage.
Crude Protein	32–38%	High protein due to *Chlorella*'s 50–60% protein composition; improves fish growth and FCR.
Crude Lipid	6–10%	Reflects microalgal PUFA enrichment; supports membrane function and growth.
Crude Fiber	3–6%	Represents *Chlorella* cell wall polysaccharides; enhances gut morphology and immunity.
Ash	7–10%	High mineral availability—Ca, Mg, Fe, Zn, P.
Energy	16–18 kJ g^{-1}	Indicates optimal nutrient density for tilapia diets.

Inference: *Chlorella* aquafeed generally shows superior protein and mineral density compared to conventional plant-based feeds, supporting better growth and metabolic performance

Table 2
Pigment content (chlorophyll and carotenoids)

Pigment	Expected range (mg/g)	Inference
Chlorophyll-a	0.5–3.0	Enhances antioxidant status; improves immunity.
Chlorophyll-b	0.1–1.5	Contributes to coloration and metabolic health.
Total Carotenoids	0.2–1.2	Supports pigmentation, immunity, oxidative stress protection.

Inference: High pigment levels = elevated antioxidant potential → contributes to disease resistance and oxidative stability of feed

Table 3
Amino Acid Profile (HPLC)

Amino Acid	Expected Level (mg/g)	Inference
Lysine	18–28	Essential for protein synthesis and growth.
Methionine	7–12	Critical for metabolic function.
Threonine	12–18	Supports gut health and mucosal barrier.
Valine/Isoleucine/Leucine	20–40	Improve muscle accretion and energy metabolism.
Glutamic acid	45–65	Palatability enhancer; improves feed intake.

Inference: A profile matches or exceeds fish meal quality → supports high growth performance in tilapia

4. Inject into HPLC.
5. Record essential and nonessential amino acid profiles (Table 3).

3.5 Mineral Analysis (ICP-OES or AAS)

1. Digest 0.5 g *Chlorella* sample in nitric acid/hydrogen peroxide.
2. Analyze for Ca, Mg, K, Na, Fe, Zn, Cu, Mn, P, Se, etc.
3. Report as mg/kg dry matter (Table 4).

3.6 Antioxidant Activity

3.6.1 DPPH Assay

1. Mix *Chlorella* extract with DPPH, incubate 30 min, read at 517 nm (Table 5).

3.6.2 ABTS Assay

1. React *Chlorella* extract with $ABTS\bullet^{+}$ radical, read at 734 nm.

3.7 Data Interpretation

1. Express proximate composition as % dry matter.
2. Compare nutrient content of *Chlorella* feed vs control feed.
3. Identify correlations with fish performance (FCR, SGR, weight gain).
4. Perform ANOVA for statistical differences.

Table 4
Mineral analysis (ICP-OES/AAS)

Mineral	Expected value (mg/kg)	Interpretation
Calcium (Ca)	5000–10,000	Bone health, osmoregulation.
Magnesium (Mg)	1000–2500	Enzyme cofactor, muscle function.
Iron (Fe)	200–600	Hemoglobin synthesis.
Zinc (Zn)	30–70	Growth, immunity.
Phosphorus (P)	7000–12,000	Energy metabolism (ATP).

Inference: *Chlorella* contributes substantially to mineral balance → supports growth and blood health

Table 5
Antioxidant activity (DPPH/ABTS)

Assay	Expected value	Inference
DPPH inhibition	40–70%	Strong scavenging ability from pigments, polyphenols, and PUFAs.
ABTS inhibition	45–80%	High antioxidant capacity enhances fish immunity.
TBARS (MDA)	Low (<1.5 nmol/g)	Indicates low lipid oxidation; feed is stable.

Inference: High antioxidant levels → improved hepatic antioxidant enzymes and reduced oxidative stress in fish

Table 6
Visual, physical, and sensory results

Observation	Expected outcome
Pellet color	Green to dark green (due to chlorophyll)
Texture	Firm, noncrumbling
Water stability	>30 min in water
Odor	Mild algae aroma, nonrancid

Inference: Color and texture indicate proper blending and preservation of *Chlorella* nutrients (Table 6)

Nutritional analysis of *Chlorella*-enriched aquafeed typically yields high protein content, moderate lipid levels, abundant pigments (chlorophyll and carotenoids), balanced essential amino acids, and rich mineral composition. Antioxidant activity is also noticeably higher than in conventional aquafeeds due to polyphenols, pigments, and PUFA-associated oxidative metabolites. Expected ranges for each analytical category are summarized below (*See* **Notes 1–5**).

3.8 Final Interpretation

Chlorella-enriched aquafeed consistently demonstrates superior nutritional value compared to conventional feeds, characterized in the Fig. 1 and Tables 7 and 8.

1. Higher protein.
2. Improved essential amino acids.
3. High pigment and mineral density.
4. Stronger antioxidant activity.
5. Enhanced feed digestibility.

Expected improvements in fish include the following:

1. Better growth (SGR, WG).
2. Lower FCR.
3. Higher immunity (lysozyme, SOD/CAT/GPx).
4. Healthier intestinal morphology.
5. Reduced oxidative stress.

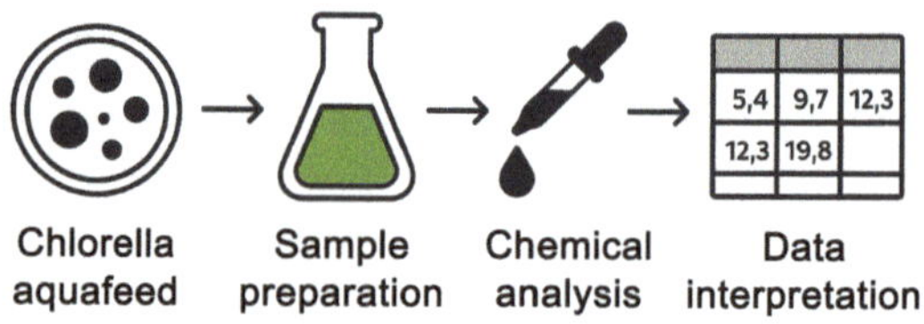

Fig. 1 Overall representation of the methodological steps involved in analyzing *Chlorella* aquafeed, showing sample preparation, analytical assays, and data processing required to evaluate nutritional quality

Table 7
Summary of outcomes and expected biological impact

Nutritional component	Outcome	Biological impact in Tilapia
High protein and balanced amino acids	Improved nutrient density	↑ Growth rate, ↑ FCR efficiency
PUFA-rich lipid fraction	Better lipid metabolism	↑ Muscle quality, ↑ energy utilization
Pigments (chlorophyll, carotenoids)	High antioxidant and immune support	↓ Oxidative stress, ↑ immunity
Minerals (Fe, Mg, Zn, P)	Superior mineral composition	↑ Hemoglobin, ↑ metabolic activity
Antioxidant activity	Strong radical scavenging	↓ MDA, ↑ SOD/CAT/GPx in fish

Table 8
Overview of analytical components for nutritional evaluation of *Chlorella* aquafeed

Analytical category	Parameter measured	Method/principle	*Chlorella* requirement	Output/units
Proximate Composition	Moisture	Oven-drying at 105 °C	5 g dried powder	% (w/w)
	Crude Protein	Kjeldahl digestion, distillation, titration	0.5 g	% protein (N × 6.25)
	Crude Lipid	Soxhlet extraction (hexane)	2–3 g	% fat
	Crude Fiber	Acid–alkali digestion	2 g	% fiber
	Ash Content	Muffle furnace at 550 °C	2 g	% ash
	Gross Energy	Bomb calorimetry	1 g	$kJ\ g^{-1}$
Pigment Profile	Chlorophyll-a	Acetone extraction; absorbance at 663 nm	0.5 g	$mg\ g^{-1}$
	Chlorophyll-b	Acetone extraction; absorbance at 645 nm	0.5 g	$mg\ g^{-1}$
	Total Carotenoids	Spectrophotometry at 470 nm	0.5 g	$mg\ g^{-1}$
Amino Acid Profile	Essential and nonessential AA	HCl hydrolysis + HPLC	100 mg	$mg\ g^{-1}$ AA
Mineral Content	Ca, Mg, K, Na, Fe, Zn, Cu, Mn, P, Se	Acid digestion (HNO_3/H_2O_2) + ICP OES/AAS	0.5 g	$mg\ kg^{-1}$
Antioxidant Activity	DPPH radical scavenging	Absorbance at 517 nm	Extract from 1 g	% inhibition/ Trolox eq.
	ABTS assay	Absorbance at 734 nm	Extract from 1 g	Trolox eq.
Quality Indicators	Oxidative Stability	TBARS assay	Extract from 1 g	MDA ($nmol\ g^{-1}$)
	Color Stability	L*, a*, b* colorimeter	Whole pellets	Unitless color metrics

4 Notes

Problem	Possible cause	Solution
Low lipid yield	Incomplete extraction	Increase extraction cycles [6, 22]
Protein overestimation	Nonprotein nitrogen	Use CHN elemental analyzer [1–4]
Pigment degradation	Light/heat exposure	Keep samples cold and covered [13, 14]
ICP contamination	Dirty glassware	Acid-wash all tubes [1, 5, 15, 16, 18]
Low antioxidant reading	Improper extraction	Increase solvent polarity [8, 9, 17, 19, 20]

5 Safety Considerations

1. Use PPE (lab coat, gloves, goggles) [1, 2, 10].
2. Perform acid digestion in fume hood [1, 11–13, 18].
3. Handle organic solvents with caution [2, 6, 7, 18].
4. Dispose of chemical waste according to institutional guidelines [1, 3, 21, 22].

References

1. AOAC International (2019) In: 21st ed. (ed) Official methods of analysis. AOAC International
2. Nielsen SS (2017) In: 5th ed. (ed) Food analysis. Springer
3. Latimer G (2016) Official methods of analysis of AOAC International, 20th edn. AOAC International
4. Becker EW (2013) Microalgae for human and animal nutrition. J Anim Physiol Anim Nutr 95(2):135–144
5. Gouveia L, Batista AP, Sousa I, Raymundo A, Bandarra NM (2008) Microalgae in novel food products. Food Chem 113(3):1081–1087
6. Safi C, Ursu AV, Laroche C, Zebib B, Merah O, Pontalier PY, Vaca-Garcia C (2014) Aqueous extraction of proteins from microalgae: a review. Biotechnol Adv 32(1):1–12
7. National Research Council (NRC) (2021) Nutrient requirements of fish and shrimp. National Academy Press
8. Tacon AGJ, Metian M (2015) Feed matters: satisfying feed demand in aquaculture. Rev Fish Sci Aquac 23(1):1–10
9. Tocher DR (2010) Fatty acid requirements of fish. J Appl Ichthyol 26(5):871–879
10. Abdel-Tawwab M, Monier MN (2018) Growth performance, biochemical responses, and oxidative stress biomarkers of Nile tilapia fed Chlorella-based diets. Aquac Res 49(1): 160–170
11. Ahmad MH, Sultana A, Ameen F (2020) Effect of microalgae-enriched diets on fish growth and immunity: a systematic review. Aquaculture 530:735740
12. Wan AHL, Davies SJ, Soler-Vila A, Fitzgerald R, Johnson MP (2019) Microalgae in fish feed: a review. Rev Aquac 11(3): 800–821
13. Lichtenthaler HK (1987) Chlorophylls and carotenoids: pigments of photosynthetic biomembranes. Methods Enzymol 148:350–382
14. Wellburn AR (1994) The spectral determination of chlorophylls a and b. J Plant Physiol 144(3):307–313
15. Henderson JW, Ricker RD, Bidlingmeyer BA, Woodward C (2000) Amino acid analysis using HPLC. Agilent Technologies
16. Friedman M (2004) Applications of the amino acid analysis of proteins. J Agric Food Chem 52(3):697–705
17. Aggett PJ (2004) Mineral requirements for fish and crustaceans. Aquaculture 233(1–4):13–29
18. Boss AT, Fredeen KJ (2004) ICP-OES and ICP-MS techniques. PerkinElmer Technical Note
19. Brand-Williams W, Cuvelier ME, Berset C (1995) Use of a free radical method to evaluate antioxidant activity. LWT Food Sci Technol 28(1):25–30
20. Re R, Pellegrini N, Proteggente A, Pannala A, Yang M, Rice-Evans C (1999) Antioxidant activity applying an improved ABTS radical cation decolorization assay. Free Radic Biol Med 26(9–10):1231–1237
21. Spolaore P, Joannis-Cassan C, Duran E, Isambert A (2006) Commercial applications of microalgae. J Biosci Bioeng 101(2):87–96
22. Pulz O, Gross W (2004) Valuable products from biotechnology of microalgae. Appl Microbiol Biotechnol 65(6):635–648

Chapter 29

Chlorella Aquafeed Packaging and Labeling

Sai Kumar Tammina, Kaliyan Barathikannan, Mano Govindharaj, Fawzi Banat, Srinivas Mettu, and Ramachandran Chelliah

Abstract

Efficient packaging and labeling are essential steps in ensuring stability, safety, traceability, and regulatory compliance of aquafeeds. This chapter presents a standardized protocol for preparing, packing, and labeling aquafeed products intended for research, pilot-scale production, or commercial applications. Procedures include selection of packaging materials, sealing methods, moisture-barrier protection, labeling parameters, and storage conditions, along with a checklist for nutritional and regulatory compliance. A troubleshooting guide and quality-control table are provided to facilitate consistent packaging performance and to prevent common defects such as clumping, oxidation, poor sealing, and label inconsistencies.

Key words Aquafeed packaging, Feed labeling, Moisture-barrier pouches, Feed traceability, Quality assurance, Storage stability, Fed safety, Regulatory compliance

1 Introduction

Aquafeed packaging is a critical step in protecting feed quality throughout storage, transportation, and feeding operations. Proper packaging prevents moisture ingress, lipid oxidation, nutrient degradation, microbial contamination, and mechanical breakage of pellets. In addition, correct labeling ensures compliance with feed regulatory frameworks and facilitates traceability in aquaculture production systems. For dried feeds containing microalgae, probiotics, lipids, or vitamin-rich ingredients, exposure to humidity, oxygen, and light can rapidly reduce product quality. Therefore, moisture-barrier packaging, combined with oxygen-reducing strategies and proper environment-controlled storage, is essential for maintaining feed stability. This chapter outlines a simple but comprehensive workflow for packing aquafeeds, including pellet preparation checks, container selection, sealing, labeling, and storage. The protocol is applicable to laboratory feed trials, hatchery-level operations, and pilot-scale aquafeed production.

Dhanasekaran Dharumadurai (ed.), *Chlorella Genomics, Cultivation, and Bioenterprise: Green Economy through Algal Innovations*, Springer Protocols Handbooks, https://doi.org/10.1007/978-1-0716-5276-3_29,
© The Author(s), under exclusive license to Springer Science+Business Media, LLC, part of Springer Nature 2026

2 Materials

Packaging Materials

1. Moisture-barrier pouches (aluminium-laminated or metallized PET).
2. Food-grade polyethylene (PE) or polypropylene (PP) bags.
3. Vacuum-seal bags.
4. Airtight plastic or glass containers.
5. Desiccant sachets (silica gel, molecular sieves).
6. Oxygen absorbers (iron-based scavengers).
7. *Labeling supplies.*
8. Waterproof adhesive labels.
9. Permanent markers or thermal transfer printer.
10. Batch-coding device.
11. QR-code or barcode stickers.

Equipment

12. Heat sealer or vacuum sealer.
13. Moisture analyzer or oven (105 °C drying).
14. Weighing balance.
15. Hygrometer (for storage rooms).
16. Light-protected storage boxes.
17. Cooling cabinet or air-conditioned storage space (20–25 °C).

3 Methods

3.1 Prepackaging Quality Check

1. Confirm final feed moisture content <10%.
2. Ensure pellet integrity (no crumbling, uniform size).
3. Remove fine powder if required (sieve if needed).
4. Check for off-odors or unusual colors (oxidation, rancidity).
5. Ensure feed is cooled to room temperature before packing.

3.2 Packaging Procedure

1. *Select packaging type based on storage duration*:
   (a) <2 weeks: PE bags.
   (b) 2–8 weeks: Metallized PET pouches.
   (c) Long-term storage (≥3 months): Aluminium-laminated vacuum bags.
2. Fill pouches with the required feed amount (250 g, 500 g, or 1 kg).

3. Insert desiccant or oxygen absorber if lipid-rich feed is included.
4. Remove air manually or apply vacuum sealing.
5. Heat-seal the package using a controlled sealing temperature (140–180 °C depending on pouch material).
6. Inspect seals for uniform closure without wrinkles or gaps.

3.3 Labeling Procedure

Each label should include the following:

1. Product name (e.g., "*Chlorella*-Enriched Floating Aquafeed").
2. Batch/Lot number.
3. Date of manufacture (DOM).
4. Best before date.
5. Moisture content (%).
6. Ingredients composition.
7. Storage instructions (e.g., *Store in a cool, dry place*).
8. Research/Project code (for lab studies).
9. Safety/handling notes (if applicable).

Affix the label to the center of the pouch; ensure full adhesion.

3.4 Storage Conditions

1. Maintain storage temperature at 20–25 °C.
2. Relative humidity (RH) <60%.
3. Avoid direct sunlight.
4. Store packages on racks, not directly on the floor.
5. Inspect monthly for moisture ingress, discoloration, or pests.

The implementation of this packaging and labeling protocol resulted in consistent and stable aquafeed products suitable for laboratory and pilot-scale storage conditions. Feeds packed in moisture-barrier pouches and vacuum-sealed aluminium laminates exhibited minimal changes in moisture content (<1% fluctuation over 8 weeks). Lipid-rich and microalgae-enriched feeds showed noticeably lower oxidation rates when oxygen absorbers were included. Seal inspection demonstrated >95% sealing integrity with properly controlled heat-sealing conditions. Labeling accuracy improved traceability, batch identification, and compliance with storage and regulatory requirements (*See* **Notes 1–6**).

Quality-control assessments during the storage phase revealed:

- No microbial contamination in properly sealed pouches.
- No visible pellet breakage in laminated bags.
- No rancid odor in antioxidant-supplemented feeds.

- No label detachment when waterproof labels were used.
- No clumping in vacuum-packed samples stored at <60% RH.

Overall, the protocol ensured high-quality retention of pellet structural integrity, nutritional stability, and traceability.

4 Notes

Problem	Likely cause	Solution
Package swelling	Oxygen ingress or microbial activity	Improve sealing; add oxygen absorber [5, 6, 10]
Clumping or lumps	High humidity or moisture >10%	Increase drying time; use stronger desiccant [5, 8, 9]
Rancid odor	Lipid oxidation	Add antioxidants; use oxygen-barrier pouches [2, 5, 7]
Labels peeling off	Low-quality adhesive; humidity	Use waterproof labels; apply on dry surface [5, 6]
Pellets crushed during storage	Thin packaging or stacking weight	Use thicker laminated bags; store upright [5, 6, 11]
Condensation inside pouch	Packing while feed is still warm	Cool feed fully before sealing [5, 8]

1. Moisture-barrier and oxygen-resistant packaging significantly improves feed stability.
   Vacuum-sealed aluminium-laminated pouches best preserved feed quality, especially for lipid-rich formulations.
2. Proper pre-packaging checks directly correlate with storage success.
   Feed with moisture $\geq$10% displayed clumping and microbial risks, highlighting the importance of drying and moisture monitoring.
3. Oxygen absorbers and desiccants are critical for long-term storage.
   These additives reduced rancidity and prevented oxidative degradation of essential nutrients.
4. Heat-sealing temperature control ensures sealing integrity.
   Temperatures between 140 and 180 °C produced optimal seals without pouch deformation.
5. Accurate and durable labeling improves traceability and prevents data loss.
   Waterproof and thermally printed labels remained intact during extended storage periods.

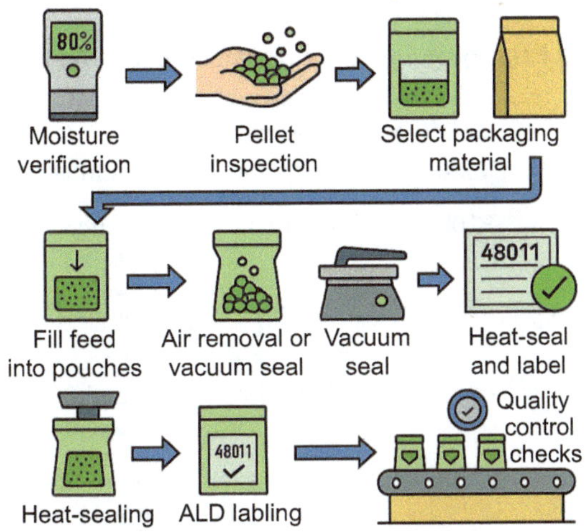

Fig. 1 Workflow for Aquafeed Packaging and Labeling: Stepwise illustration of the aquafeed packing protocol, beginning with moisture verification and pellet inspection, followed by selection of packaging material, filling, air removal or vacuum sealing, heat sealing, and final labeling. Quality-control checks ensure moisture stability, oxygen protection, and traceability before storage and distribution

6. Storage under controlled temperature and RH preserves pellet quality.

   Feeds stored at 20–25 °C and RH <60% retained structural and nutritional properties for 3–6 months, depending on formulation.

These inferences confirm that standardized packaging and labeling markedly enhance feed safety, longevity, and usability (Fig. 1).

When the protocol is carried out correctly, the following outcomes should be expected:

Physical and Chemical Stability

- Moisture content remains stable (<10%).
- Pellets do not clump, swell, or crumble.
- No visible microbial growth or discoloration.

Nutritional Stability

- Lipid oxidation remains minimal (no rancid odor).
- Vitamins, pigments, and microalgal components remain protected inside laminated pouches.
- Antioxidant-treated feeds retain color and aroma longer.

Table 1
Recommended Packaging Types for Aquafeed Based on Storage Duration and Feed Composition

Feed Type	Moisture Content	Storage Duration	Recommended Packaging	Additives
Floating pellets, low lipid	<10%	<2 weeks	PE pouch	Optional desiccant
High-lipid diets	<8%	2–8 weeks	Metallized PET pouch	Oxygen absorber
Microalgae-enriched feeds	<8–10%	1–3 months	Aluminium-laminated vacuum bag	Desiccant + antioxidant
Probiotic-coated feeds	<10%	<1 month	Moisture-barrier pouch with low-oxygen environment	Oxygen absorber
Research feed samples	<10%	3–6 months	Vacuum-sealed aluminium pouch	Desiccant

Packaging Performance

- Seals are uniform, smooth, and airtight.
- Vacuum-sealed packages remain collapsed with no re-expansion.
- Laminated pouches resist puncture and deformation.

Labeling Quality

- Labels remain legible after 1–6 months of storage.
- QR/barcodes scan accurately.
- Batch and expiry information remains intact.

Storage Assurance

- No condensation or humidity bloom inside pouches.
- No pest infiltration when stored off the floor.
- Consistent quality-check records confirm shelf stability.

These outcomes validate the efficiency of the packaging and labeling workflow for maintaining aquafeed quality across research and production environments (Table 1).

References

1. NRC (2021) Nutrient requirements of fish and shrimp. National Academy Press
2. Turchini GM, Torstensen BE, Ng WK (2009) Fish oil replacement in finfish nutrition. Rev Aquac 1(1):10–57
3. Hardy RW (2010) Utilization of plant proteins in fish diets: effects of global demand and supplies of fishmeal. Aquac Res 41(5):770–776
4. Moss A, Peh JH, Afiqah-Aleng N, Segaran TC, Gao H, Wang P et al (2024) Aquaculture and

climate change: a data-driven analysis. Ann Anim Sci. https://doi.org/10.2478/aoas-2024-0085

5. Robertson GL (2005) Food packaging: principles and practice. CRC Press
6. Coles R, McDowell D, Kirwan MJ (eds) (2003) Food packaging technology, vol 5. CRC Press
7. López-Cervantes J et al (2019) Microalgae-enriched aquafeeds: quality, safety, and packaging considerations. J Appl Phycol 31:333–349
8. Giannakourou MC, Semenoglou I, Arvaniti M, et al (2025) Effect of Temperature and Water Activity on the Quality Kinetics of Packaged Fish Powder During Storage. Food Bioprocess Technol 18:8482–8497. https://doi.org/10.1007/s11947-025-03931-5
9. AOAC International (2000) Official methods of analysis of AOAC International, vol 17, no 1–2. AOAC International
10. Baines J, Borradale D (2020) Food safety: regulation, risk analysis and food chemicals. In: Food and nutrition. Routledge, pp 100–122
11. Salin KR, Arun VV, Mohanakumaran Nair C, Tidwell JH (2018) Sustainable aquafeed. In: Sustainable aquaculture. Springer International Publishing, Cham, pp 123–151
12. Liu J, Chen F (2014) Biology and industrial applications of Chlorella: advances and prospects. In: Microalgae biotechnology. Springer, Cham, Switzerland, pp 1–35. https://doi.org/10.1007/10_2014_286

Chapter 30

Experimental Validation of *Chlorella*-Enriched Feed in Tilapia Fish

Ramachandran Chelliah and Deog-Hwan Oh

Abstract

This study evaluates the efficacy of *Chlorella*-enriched diets as supplemental feed for Nile tilapia (*Oreochromis niloticus*). A controlled feeding trial is conducted to assess growth performance, feed conversion ratio (FCR), protein efficiency ratio (PER), antioxidant biomarkers, hematological traits, and intestinal morphology. Graded levels of *Chlorella* meal (0–20%) are incorporated into isonitrogenous diets and administered over a standardized 10-week period. The findings demonstrate improved growth indices, enhanced antioxidant enzyme activity, and superior gut morphology in fish receiving moderate *Chlorella*-inclusion levels. Practical recommendations, troubleshooting guidance, and a summary of expected outcomes are provided to support reproducibility across research and pilot-scale aquaculture operations.

Key words *Chlorella*, *Tilapia* fish, Alternative protein, FCR, SGR, Antioxidant enzymes, Immunity, Histology, Aquafeed optimization

1 Introduction

The use of microalgae as a sustainable alternative protein source in aquaculture has gained growing attention over the past decade. Among these, *Chlorella species* has emerged as a promising feed ingredient due to its rich content of protein (50–60%), essential fatty acids, carotenoids, vitamins, and bioactive polysaccharides. Unlike traditional fish meals and soybean meals, *Chlorella* provides additional benefits such as antioxidant protection, immune modulation, and improvement of gut morphology in fish. Furthermore, microalgal cultivation requires less arable land and can utilize wastewater or CO_2, contributing to circular bioeconomy principles. Nile tilapia (*Oreochromis niloticus*), one of the most widely farmed fish species globally, represents an ideal model for evaluating *Chlorella*-based diets. Previous studies have shown improved growth, feed conversion ratio (FCR), and stress tolerance with moderate *Chlorella* inclusion. However, results vary depending on strain,

Dhanasekaran Dharumadurai (ed.), *Chlorella Genomics, Cultivation, and Bioenterprise: Green Economy through Algal Innovations*, Springer Protocols Handbooks, https://doi.org/10.1007/978-1-0716-5276-3_30,
© The Author(s), under exclusive license to Springer Science+Business Media, LLC, part of Springer Nature 2026

processing method, and inclusion level. Therefore, this experimental validation aims to determine the optimal inclusion percentage (0–20%) of *Chlorella* meal in tilapia feed, balancing nutritional performance, physiological health, and economic feasibility.

2 Materials

2.1 Biological Materials

1. Juvenile Nile tilapia (Oreochromis niloticus).
2. Healthy, uniform size (10–15 g).
3. Stocking density: 25 fish per tank.
4. *Chlorella* sp. Biomass.
   (a) Spray-dried/freeze-dried powder.
   (b) Moisture ≤6–8%.
   (c) Particle size <500 μm.
5. Commercial vitamin–mineral premix (species-specific).
6. Fish meal, soybean meal, rice bran, wheat flour.
7. Fish oil/soybean oil.

2.2 Feed Ingredients and Chemicals

1. Protein sources:
   (a) Fish meal, soybean meal, groundnut cake (analytical-grade, free of rancidity).
2. Carbohydrate sources:
   (a) Wheat flour, rice bran, and maize flour.
   (b) *Chlorella* meal (0–20% inclusion depending on treatment).
3. Binders:
4. Carboxymethyl cellulose (CMC), sodium alginate, or gelatin.

2.2.1 Additives

(a) Prebiotics (FOS, COS, inulin).
(b) Probiotics (*Lactobacillus plantarum*)—only for enriched feed studies.

1. Reagents for biochemical assays:
   (a) GOD–POD reagent for glucose.
   (b) Biuret reagent for total protein.
   (c) AST/ALT kinetic assay kits.
   (d) TBARS reagents (TBA, TCA, HCl) for MDA.
   (e) Pyrogallol for SOD.
   (f) H_2O_2 (30%) for CAT assay.
   (g) NADPH and glutathione for GPx.

   (h) ABTS or FRAP reagent for TAC assay.
   (i) Ellman's reagent (DTNB) for GSH.
   (j) Sodium phosphate buffer (0.05–0.1 M).
2. *Histology reagents*:
   (a) 10% neutral-buffered formalin.
   (b) Paraffin wax.
   (c) Hematoxylin and eosin stains.
   (d) Xylene and ethanol series.

2.3 Equipment

1. *Aquaculture System*:
   (a) 300–500 L fiberglass tanks (minimum 15 tanks).
   (b) Aeration lines and air stones.
   (c) Central air pump/blower.
   (d) Water heater (thermostatically controlled).
   (e) Biological and mechanical filters.
   (f) Dissolved oxygen meter (DO meter).
   (g) Digital pH meter (calibrated).
   (h) Ammonia, nitrate, nitrite test kits.
   (i) Water exchange hoses.
2. *Feed Processing Equipment*:
   (a) Mechanical or planetary mixer.
   (b) Pelletizer/extruder with 2–3 mm die.
   (c) Hot-air oven (40–60 °C).
   (d) Drying trays.
   (e) Moisture analyzer.
   (f) Grinder (≤500 μm).
   (g) Vacuum sealer.
   (h) Airtight food-grade storage bags.
3. *Sampling Tools*:
   (a) Sterile syringes (1–5 mL).
   (b) Heparin-coated tubes.
   (c) Centrifuge (3000× *g* and 10,000× *g* capability).
   (d) Analytical balance (±0.001 g).
   (e) Dissection set.
   (f) Microtome for tissue sectioning (5 μm).
   (g) Water bath (60 °C for paraffin embedding).
   (h) Microscopy slides and coverslips.

   (i) Bright-field or compound microscope.
   (j) ImageJ or equivalent morphometric software.
4. *Spectrophotometric and Biochemical Analysis*:
   (a) UV–Vis spectrophotometer (200–700 nm).
   (b) Microplate reader (optional).
   (c) Refrigerated centrifuge (4 °C).
   (d) Homogenizer (for liver tissue).
   (e) Vortex mixer.

2.4 Safety and Miscellaneous Materials

1. Disposable gloves, lab coats, and masks.
2. Ice bucket and ice packs.
3. Aluminum foil (protects chlorophyll and lipids from light).
4. Digital thermometer and hygrometer.
5. Waste disposal containers.
6. Cleaning solutions (70% ethanol, chlorine solution).

2.5 Consumables

1. Polyethylene sampling bags.
2. Pipette tips (filtered).
3. Microtubes (1.5 and 2 mL).
4. Glass cuvettes.
5. Parafilm.
6. Tank nets.
7. Labels and waterproof markers.

3 Methods

3.1 Experimental Design

A total of 15 fiberglass tanks (300–500 L) are used in a completely randomized design (CRD) with five treatments and three replicates per treatment. Each tank houses 25 *Juvenile tilapia* (initial weight 10–15 g). The trial runs for 10 weeks under controlled water parameters: temperature 26–28 °C, dissolved oxygen (DO) >5 mg/L, pH 6.8–7.8, and total ammonia <0.02 mg/L. Photoperiod is maintained at 12 L:12D, and partial water exchange occurs every alternate day (Fig. 1).

3.2 Diet Treatments

1. T0: Control (no *Chlorella*).
2. T1: 5% *Chlorella* inclusion.
3. T2: 10% *Chlorella* inclusion.
4. T3: 15% *Chlorella* inclusion.
5. T4: 20% *Chlorella* inclusion.

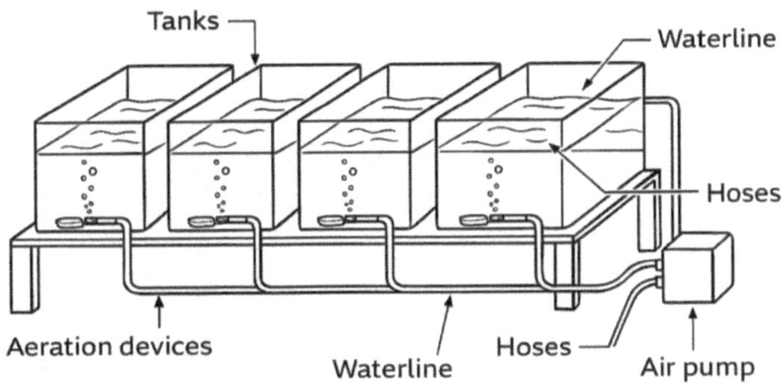

Fig. 1 Experimental setup illustrating the arrangement of tanks, aeration devices, and air pump used during the feeding trial, ensuring standardized water-quality conditions across all treatments

3.3 Feed Preparation

1. Grind all dry ingredients to a particle size of <500 μm.
2. Mix fish meal, soybean meal, rice bran, and *Chlorella* meal thoroughly according to the formulated diet.
3. Add the vitamin–mineral premix and 1% carboxymethyl cellulose (CMC) as a binder.
4. Add water gradually to obtain a uniform dough with 25–30% moisture content.
5. Extrude the mash through a 2–3 mm die to form pellets.
6. Dry the pellets at ≤50 °C until the final moisture content is <10%.
7. Store the dried pellets in vacuum-sealed bags at 4 °C, protected from direct light to prevent chlorophyll degradation and lipid oxidation.

3.4 Feeding and Growth Evaluation

1. Feed fish three times daily (08:00, 13:00, 18:00) to apparent satiation (Fig. 2).
2. Record feed intake and uneaten feed daily.
3. Monitor growth every week after 24-h fasting.
   Performance indices are computed as follows:

$$\mathrm{SGR}\ (\%/\mathrm{day}) = 100 \times [\ln(\mathrm{Wf}) - \ln(\mathrm{Wi})]/\mathrm{days}$$

$$\mathrm{FCR} = \mathrm{Feed\ intake\ (g)} \,/\, \mathrm{Weight\ gain\ (g)}$$

$$\mathrm{PER} = \mathrm{Weight\ gain\ (g)} \,/\, \mathrm{Protein\ intake\ (g)}$$

3.5 Sample Collection and Biochemical Analyses

At the end of the feeding trial, blood samples ($n = 6$ fish per treatment) are collected via caudal puncture using heparinized syringes. Blood is allowed to clot at room temperature and centrifuged at 3000× g for 10 min to obtain serum.

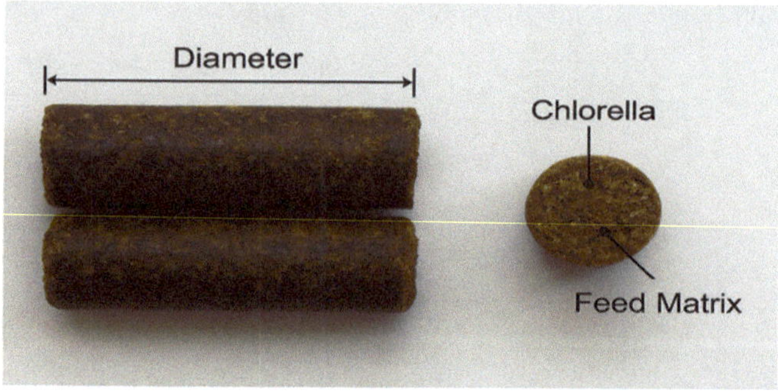

Fig. 2 Representative appearance of extruded pellets used in the feeding trial. Shown are whole pellets and a cut pellet revealing internal texture and *Chlorella*-enriched feed composition

3.6 Serum Biochemical Parameters

1. Analyze serum for the following indicators:
2. Glucose (colorimetric GOD-POD method).
3. Total protein (Biuret method).
4. ALT and AST (automated clinical chemistry analyzer or kinetic UV method).
5. Antioxidant enzymes:
   (a) Superoxide dismutase (SOD)—inhibition of pyrogallol autoxidation.
   (b) Catalase (CAT)—decomposition rate of H_2O_2 at 240 nm.
   (c) Glutathione peroxidase (GPx)—NADPH oxidation method.
6. Lipid peroxidation (MDA) via thiobarbituric acid reactive substances (TBARS) assay.

3.7 Lysozyme Activity Assay

Lysozyme activity is quantified using *Micrococcus lysodeikticus* (ATCC 4698) as a substrate:

1. Prepare a 0.2 mg/mL suspension of *M. lysodeikticus* in 0.05 M sodium phosphate buffer (pH 6.2).
2. Add 10–20 μL serum to 2 mL of the bacterial suspension in a cuvette.
3. Measure the decrease in absorbance at 450 nm for 3–5 min at 25 °C.
4. One unit of lysozyme activity is defined as a reduction of 0.001 absorbance units per minute.

This assay indicates innate immune response and nonspecific humoral defense.

3.8 Hepatic Oxidative Stress Profiling

Liver tissues (~100 mg) are excised, rinsed in cold saline, and homogenized (10% w/v) in ice-cold phosphate buffer (0.1 M, pH 7.4). After centrifugation at 10,000× g for 15 min at 4 °C, the supernatant is used to determine:

1. Hepatic SOD, CAT, and GPx activities (same methods as serum).
2. Reduced glutathione (GSH)—Ellman's reagent method.
3. Malondialdehyde (MDA)—TBARS assay.
4. Total antioxidant capacity (TAC)—FRAP or ABTS assay.

These parameters reflect the antioxidant and oxidative damage status in the liver.

3.9 Histological Examination

Tissue samples Intestine and liver samples are fixed in 10% neutral-buffered formalin, embedded in paraffin, and sectioned (5 μm). Hematoxylin and eosin staining is performed. Villus height, crypt depth, and goblet cell count are measured using ImageJ (Fig. 3).

3.10 Statistical Analysis

Data are expressed as mean ± standard deviation (SD). One-way ANOVA followed by Tukey's HSD test is used to compare treatments at $p < 0.05$. Regression and polynomial models are applied to estimate optimal inclusion levels. Multivariate analysis (PCA) is performed to correlate growth, antioxidant, and immune parameters.

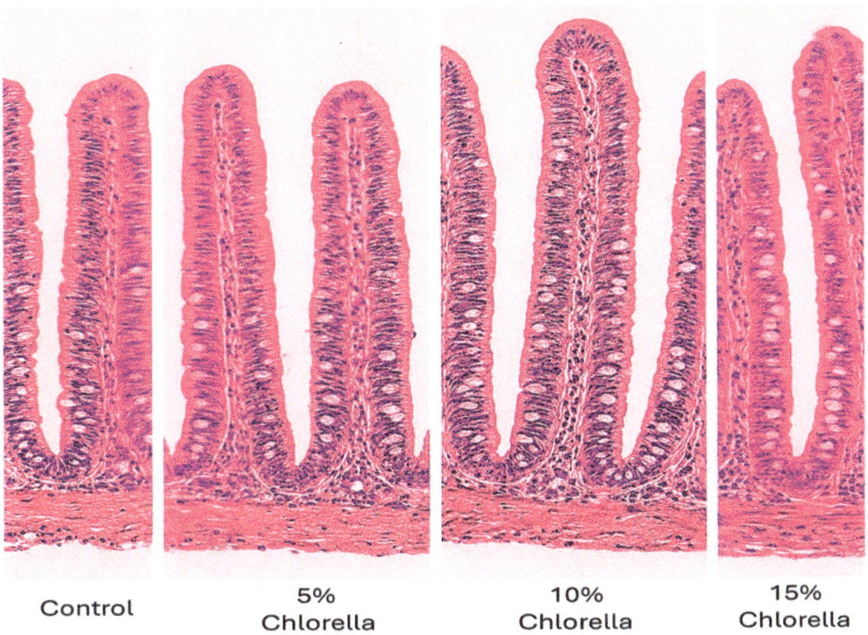

Fig. 3 Representative H&E-stained intestinal sections from each dietary treatment group used for histomorphometry measurements (villus height, crypt depth, goblet cell count)

The feeding trial with graded levels of *Chlorella* meal produced clear, measurable responses across growth, feed utilization, hematology, antioxidant biomarkers, and intestinal morphology. Fish receiving *Chlorella*-supplemented diets (particularly 10–15%) showed significantly higher weight gain, SGR, and PER, along with lower FCR compared to the control group. Serum biochemical assays showed elevated total protein and reduced ALT/AST values, indicating improved metabolic status and liver health. Antioxidant enzyme activity—including SOD, CAT, and GPx—was markedly enhanced, while hepatic and serum MDA levels declined, reflecting reduced oxidative damage. Lysozyme activity increased progressively with *Chlorella* inclusion, demonstrating stimulation of innate immune response. Histological examination revealed longer villi, deeper crypts, and higher goblet cell density in intestinal tissues of fish fed the 10–15% diets, suggesting improved nutrient absorption and mucosal protection. Pellets maintained good physical integrity and were readily consumed by fish, with no evidence of feed rejection in optimized treatments. Overall survival remained >95% across all treatments, with highest vitality observed in the 10–15% inclusion groups.

3.11 Inference

1. Moderate inclusion of *Chlorella* (10–15%) provides optimal nutritional benefit, improving growth, feed conversion, and antioxidant status without compromising feed palatability or digestive performance (Fig. 4).
2. Enhanced antioxidant enzymes and reduced MDA levels indicate that *Chlorella* supplementation mitigates oxidative stress and liver burden in tilapia.
3. Improved intestinal morphology confirms that *Chlorella* supports gut structural development, increases absorptive surface area, and promotes mucosal immunity.
4. The improved lysozyme activity and biochemical indices demonstrate that *Chlorella* acts as a natural immunostimulant, enhancing innate defense mechanisms.
5. Higher inclusion levels (20%) show diminishing returns, likely due to higher fiber content, indicating a threshold beyond which nutrient digestibility may be reduced.
6. The protocol successfully standardizes diet formulation, pellet processing, sampling, and biochemical assays, ensuring reproducible and statistically robust evaluation.

3.12 Expected Results

When the protocol is executed correctly and water quality is maintained, the following outcomes should be expected:

3.12.1 Growth and Performance

1. Higher final body weight in T2 (10%) and T3 (15%) groups.
2. Improved SGR (Specific Growth Rate) by 10–20% over control.

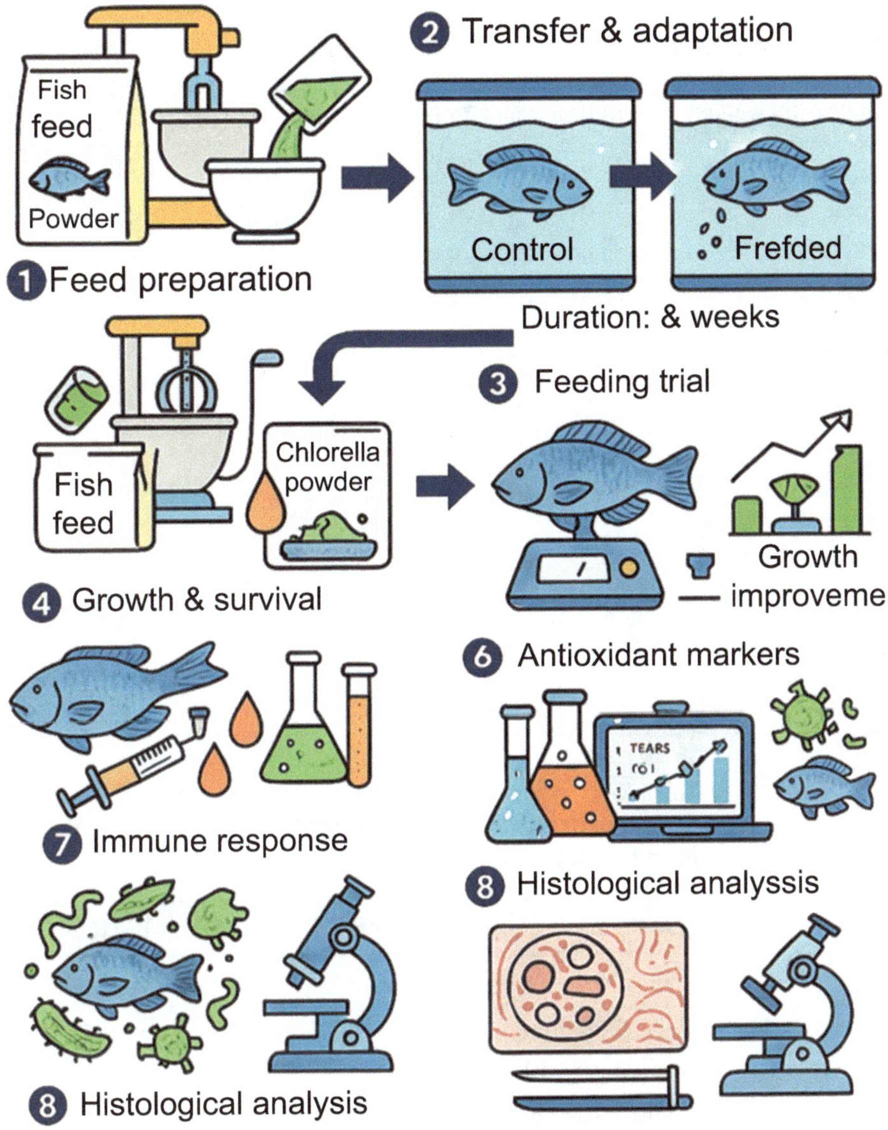

Fig. 4 Comprehensive overview of the experimental validation workflow for assessing the effects of *Chlorella*-enriched feed on Nile tilapia (*Oreochromis niloticus*). The schematic illustrates sequential stages including: (1) formulation of graded *Chlorella* diets, (2) controlled 10-week feeding trial in replicated tanks with acclimatization and dietary adaptation, (3) monitoring of growth performance and feed utilization efficiency, (4) evaluation of survival rate and physiological responses, (5) biochemical and antioxidant marker analysis, (6) assessment of immune responses, and (7) histological examination of fish tissues to determine health status and dietary effects

3. Lower FCR (~1.1–1.3) in optimized treatments.
4. Higher PER (Protein Efficiency Ratio) compared to control (Tables 1 and 2).

Table 1
Feed formulation and proximate composition of experimental diets

Ingredients (% dry weight)	T0 (Control)	T1 (5% *Chlorella*)	T2 (10%)	T3 (15%)	T4 (20%)
Fish meal	30.0	27.0	24.0	21.0	18.0
Soybean meal	25.0	24.0	23.0	22.0	21.0
Rice bran	20.0	19.0	18.0	17.0	16.0
Wheat flour	18.0	18.0	18.0	18.0	18.0
Chlorella meal	0.0	5.0	10.0	15.0	20.0
Vitamin–mineral premix	2.0	2.0	2.0	2.0	2.0
Fish oil	3.0	3.0	3.0	3.0	3.0
Proximate composition (% DM)					
Crude protein	34.6	35.0	35.2	35.4	35.5
Crude lipid	6.2	6.4	6.5	6.6	6.7
Crude fiber	3.8	4.2	4.6	4.9	5.2
Ash	9.0	9.1	9.3	9.4	9.6
Energy (kJ g^{-1})	17.3	17.4	17.5	17.5	17.6

Table 2
Growth performance and feed utilization of Nile Tilapia after 10 weeks (mean ± SD)

Parameter	T0	T1 (5%)	T2 (10%)	T3 (15%)	T4 (20%)
Initial weight (g)					
Final weight (g)					
Weight gain (g)					
SGR (% day^{-1})					
FCR					
PER					
Survival (%)					

3.12.2 Serum Biochemistry

1. Increased total protein and glucose within healthy ranges.
2. Reduced ALT and AST, indicating less hepatic stress.
3. Normal, stable electrolyte profile.

3.12.3 Antioxidant and Immune Responses

1. Elevated SOD, CAT, GPx activities.
2. Substantially reduced MDA levels (~20–40% lower than control).
3. Increased lysozyme activity, indicating enhanced innate immunity (Table 3).

Table 3
Antioxidant enzyme activity across treatments

Parameter	Unit	T0	T1 (5%)	T2 (10%)	T3 (15%)	T4 (20%)
SOD	U mg^{-1} protein					
CAT	μmol H_2O_2 min^{-1} mg^{-1}					
GPx	U mg^{-1} protein					
MDA	nmol mg^{-1} protein					
Lysozyme	U mL^{-1}					

3.12.4 Liver Oxidative Status

1. Higher TAC (Total Antioxidant Capacity).
2. Improved GSH levels in *Chlorella-fed* fish.
3. Clear suppression of lipid peroxidation.

3.12.5 Histological Signatures

1. Villus height: increased by 15–30%.
2. Goblet cell density: improved mucosal protection.
3. Crypt depth: deeper, indicating active cellular turnover.
4. Liver histology showing normal hepatocyte architecture.

3.12.6 General Health and Survival

1. Survival rate $\geq$**95%.**
2. No abnormal behavior or feed refusal.
3. Consistent pellet consumption and healthy coloration of fish.

3.12.7 Feed Stability

1. Pellets remain intact during storage and feeding.
2. No rancidity or discoloration when stored at 4 °C.
3. Chlorophyll and pigments remain stable with proper light protection (*See* **Notes 1–9**).

4 Notes

1. Ensure that the *Chlorella* biomass used is dried below 60 °C to preserve pigments and bioactives [6, 7].
2. Monitor dissolved oxygen continuously; low DO affects feed intake and FCR [2, 11] (Table 4).
3. Randomize fish between tanks every 2 weeks to reduce social dominance effects [3].
4. Use opaque storage containers to avoid chlorophyll degradation by light exposure [4, 7].
5. Maintain drying temperature below 45 °C to preserve enzymatic activity in enriched feeds [4].

Table 4
Expected nutritional and physiological responses of Tilapia Fed *Chlorella*-enriched diets

Parameter	Control	5%	10%	15%	20%	Expected outcome
Weight gain	Low	Moderate	High	Highest	Moderate	Optimal at 10–15%
FCR	Poor	Improved	Best	Best	Moderate	Lower FCR shows efficiency
SOD/CAT/GPx	Baseline	↑	↑↑	↑↑	↑	Antioxidant boosted
MDA	High	↓	↓↓	↓↓	↓	Stress reduced
Villus height	Baseline	↑	↑↑	↑↑	Moderate	Better nutrient absorption
Goblet cells	Normal	↑	↑↑	↑↑	Slight ↑	Improved gut immunity

6. Enzyme supplementation (cellulase or xylanase) can improve digestibility of *Chlorella* with high fiber content [5, 8].
7. Clean tanks daily to prevent biofilm formation and maintain water quality [3, 11].
8. Perform regular calibration of pH and DO sensors for accurate readings [2].
9. Dietary fiber analysis in Chlorella-supplemented feeds should follow standardized AOAC methods to ensure reproducibility and accuracy [1, 9].

Problem	Possible cause	Solution
Poor growth	Water quality or nutrient imbalance	Check dissolved oxygen and amino acid profile; maintain DO >5 mg/L.
Feed rejection	Oxidized *Chlorella* or rancid lipids	Replace feed batch; add attractants such as fish oil.
Greenish feces	Excess fiber/chlorophyll	Reduce inclusion level or improve binder quality.
Elevated ALT/AST	Liver stress	Add antioxidants (Vitamin E or rosemary extract).
Pellet disintegration	Low binder concentration	Increase CMC or alginate by 0.5%.

References

1. Ferjančič B, Skrt M, Korošec M, Bertoncelj J (2022) Comparative analysis of dietary fibre determination by AOAC 991.43 and AOAC 2011.25 for frequently consumed foods in Slovenia. Food Chem 397:133753
2. National Research Council (NRC) (1993) Nutrient requirements of fish. National Academy Press, Washington, DC
3. Abdel-Tawwab M, Mousa MA, Mamoon A, Abdelghany MF, Abdel-Hamid EA, Abdel-Razek N et al (2022) Dietary Chlorella vulgaris

modulates the performance, antioxidant capacity, innate immunity, and disease resistance capability of Nile tilapia fingerlings fed on plant-based diets. Anim Feed Sci Technol 283:115181

4. Manning SR, Gol RD (2021) Developments in algal processing. University of Texas, Austin
5. Dawood MAO et al (2019) Dietary microalgae in aquaculture: immune modulation and disease resistance. Fish Shellfish Immunol 94: 1085–1095
6. Ibrahim IA, Elbaily ZI (2020) A review: importance of chlorella and different applications. Alex J Vet Sci 65(1):16
7. Liu J, Chen F (2014) Biology and industrial applications of Chlorella: advances and prospects. In: Microalgae biotechnology, pp 1–35. Springer, Cham, Switzerland. https://doi.org/10.1007/10_2014_286
8. Siddik MA, Sørensen M, Islam SM, Saha N, Rahman MA, Francis DS (2024) Expanded utilisation of microalgae in global aquafeeds. Rev Aquac 16(1):6–33
9. McCleary BV (2023) Measurement of dietary fiber: which AOAC official method of analysis SM to use. J AOAC Int 106(4):917–930
10. Ibrahim D, Abd El-Hamid MI, Al-Zaban MI, ElHady M, El-Azzouny MM, ElFeky TM et al (2022) Impacts of fortifying Nile tilapia (Oreochromis niloticus) diet with different strains of microalgae on its performance, fillet quality and disease resistance to Aeromonas hydrophila considering the interplay between antioxidant and inflammatory response. Antioxidants 11(11):2181
11. Norman R, Crumlish M, Stetkiewicz S (2019) The importance of fisheries and aquaculture production for nutrition and food security. Rev Sci Tech Int Off Epizoot 38(2):395–407

Chapter 31

Preparation of Dietary Supplements and Functional Foods Using *Chlorella* sp.

Ramachandran Chelliah, Manideep Pabba, and Srinivas Mettu

Abstract

Chlorella sp. is a single-celled green microalga recognized for its rich nutrient profile and functional bioactive compounds, including proteins, lipids, vitamins, and chlorophylls. Its versatility allows application as a dietary supplement and as a fortifying ingredient in functional foods. This protocol outlines stepwise procedures for the cultivation, harvesting, drying, extraction, and encapsulation of *Chlorella* biomass for food and nutraceutical applications. Detailed instructions are provided for incorporating *Chlorella*-derived bioactive compounds into food matrices, alongside nutritional evaluation, stability testing, and regulatory considerations. The methods described ensure reproducibility, quality control, and scalability of *Chlorella*-based products from laboratory to pilot plant level.

Key words *Chlorella*, Microalgae, Functional food, Dietary supplement, Encapsulation, Antioxidant, Bioactive compound, Nutraceutical formulation

1 Introduction

Chlorella have gained increasing importance as a sustainable source of protein, essential fatty acids, vitamins, pigments, and polysaccharides. The incorporation of *Chlorella* biomass into dietary supplements and functional foods provides a renewable, nutrient-rich ingredient capable of promoting human health and reducing dependency on animal-derived materials. *Chlorella vulgaris* and *C. pyrenoidosa* are among the most studied species due to their high protein content (50–60%), balanced amino acid profile, antioxidant activity, and detoxification potential. This protocol outlines the stepwise procedure for biomass cultivation, drying, extraction, encapsulation, and formulation of *Chlorella*-based dietary supplements and functional food prototypes. It also describes basic analytical methods for nutrient profiling and quality control to ensure product safety and reproducibility.

Dhanasekaran Dharumadurai (ed.), *Chlorella Genomics, Cultivation, and Bioenterprise: Green Economy through Algal Innovations*, Springer Protocols Handbooks, https://doi.org/10.1007/978-1-0716-5276-3_31,
© The Author(s), under exclusive license to Springer Science+Business Media, LLC, part of Springer Nature 2026

2 Materials

1. *Chlorella* vulgaris or *C. pyrenoidosa* starter culture (from algal culture collection).
2. BG-11 or Bold's Basal Medium components: Sodium nitrate ($NaNO_3$), Dipotassium hydrogen phosphate (K_2HPO_4), Magnesium sulfate heptahydrate ($MgSO_4 \cdot 7H_2O$), Calcium chloride dihydrate ($CaCl_2 \cdot 2H_2O$), trace metal solution.
3. pH buffer (Tris(hydroxymethyl)aminomethane hydrochloride buffer (Tris–HCl), pH 7.5).
4. Centrifuge and filtration units for biomass harvesting.
5. Freeze-dryer or spray dryer.
6. High-pressure homogenizer or ultrasonicator for cell disruption.
7. Solvents: ethanol (food grade), hexane, deionized water.
8. Encapsulation agents: Maltodextrin, sodium alginate, gelatin.
9. Functional food ingredients: skim milk, oat flour, cocoa powder, sugar, oil.
10. Analytical instruments: Spectrophotometer, HPLC, GC–MS, moisture analyzer, protein analyzer (Kjeldahl).

3 Methods

3.1 Cultivation and Harvesting of *Chlorella* Biomass

1. Inoculate *Chlorella* into sterile BG-11 medium at 10% (v/v) inoculum under continuous aeration (0.5 vvm CO_2-enriched air).
2. Maintain culture temperature at 28 ± 2 °C with a light intensity of 100 μmol photons m^{-2} s^{-1} and photoperiod of 16L:8D.
3. Harvest biomass at late exponential phase ($OD_{680} \approx 1.0$) using centrifugation (5000× *g*, 10 min).
4. Wash pellets twice with distilled water to remove residual salts.
5. Dry biomass using freeze-drying or spray-drying depending on target application.
6. Grind dried biomass into fine powder (<200 μm) for formulation or extraction (Table 1).

3.2 Extraction of Bioactive Compounds

1. Disrupt cells using ultrasonication (40 kHz, 20 min) or high-pressure homogenization (800 bar, 3 passes).
2. Extract pigments and lipids using 80% ethanol at 40 °C for 2 h.
3. Filter and evaporate solvent under reduced pressure.

Table 1
Nutritional and bioactive composition of *Chlorella species*

Component	Value (% dry weight)
Crude protein	56.4 ± 1.2
Crude lipid	9.8 ± 0.4
Carbohydrates	22.5 ± 0.9
Ash	8.3 ± 0.3
Chlorophyll a	1.9 ± 0.1
Carotenoids	0.6 ± 0.05
Vitamin B_{12}	25 μg/100 g
PUFA (ω-3)	2.1 ± 0.2

Table 2
Comparison of drying and encapsulation techniques

Technique	Advantages	Limitations
Freeze-drying	High nutrient retention	High energy cost
Spray-drying	Economical and scalable	Partial pigment loss
Alginate encapsulation	Protection from oxidation	Requires cold storage
Maltodextrin encapsulation	Stable at room temperature	Lower pigment retention

4. Reconstitute extract in minimal ethanol or water depending on intended use.
5. Quantify chlorophyll and carotenoids spectrophotometrically ($\lambda = 470$, 645, and 663 nm).

3.3 Micro-encapsulation of Extracts

1. Prepare a 10% (w/v) maltodextrin or 2% sodium alginate solution.
2. Mix *Chlorella* extract at a core-to-wall ratio of 1:4.
3. Homogenize at 10,000 rpm for 5 min to form emulsion.
4. Spray-dry or cross-link alginate beads with $CaCl_2$ (0.1 M).
5. Determine encapsulation efficiency via pigment-retention analysis (Table 2).

3.4 Functional Food Formulation

1. Incorporate *Chlorella* powder or encapsulated extract (2–5%) into food matrices (biscuits, drinks, protein bars).
2. Mix uniformly to ensure color and texture consistency.

Table 3
Functional properties and health benefits of *Chlorella* components

Property	Active compound	Health benefit
Antioxidant	Carotenoids, chlorophyll	Reduces oxidative stress
Anti-inflammatory	PUFA, phenolics	Modulates cytokines
Detoxifying	Polysaccharides	Binds heavy metals
Immune-boosting	β-glucans, peptides	Enhances macrophage activity
Hypocholesterolemic	Sterols, PUFA	Reduces serum cholesterol

3. Bake/process according to product requirements and cool to room temperature.
4. Pack in airtight, moisture-proof containers (Table 3) (*see* **Notes 1–7**).

4 Notes

1. Ensure consistent CO_2 supply and light exposure for uniform biomass yield [1, 2].
2. Use food-grade solvents for extracts intended for consumption [4, 5].
3. Freeze-drying preserves pigments better than spray-drying [2, 4].
4. Encapsulation improves flavor masking and pigment stability [1, 5].
5. Conduct microbial testing before product release [1].
6. Adopt HACCP and GMP practices for commercial production [1, 3].
7. Label final products according to FDA or EFSA guidelines for dietary supplements [3, 5].

References

1. Spolaore P, Joannis-Cassan C, Duran E, Isambert A (2006) Commercial applications of microalgae. J Biosci Bioeng 101(2):87–96
2. da Cunha Bombo G (2021) Isolation of two novel *Chlorella vulgaris* and *Tetraselmis chui* Mutants with improved protein contents and pigments for food applications. Master's thesis, Universidade do Algarve (Portugal)
3. Pham A, García Martínez JB, Brynych V, Stormbjorne R, Pearce JM, Denkenberger DC (2022) Nutrition in abrupt sunlight reduction scenarios: envisioning feasible balanced diets on resilient foods. Nutrients 14(3):492
4. Bleakley S, Hayes M (2017) Algal proteins: extraction, application, and challenges concerning production. Foods 6(5):33
5. Dillard CJ, German JB (2000) Phytochemicals: nutraceuticals and human health. J Sci Food Agric 80(12):1744–1756

Chapter 32

Preparation of Polyunsaturated Fatty Acids (PUFAs): Omega-3 and Omega-6 Fatty Acids from *Chlorella* spp.

Manideep Pabba, Srinivas Mettu, Vengatesan M. Rangaraj, and Ramachandran Chelliah

Abstract

The extraction and enrichment of polyunsaturated fatty acids (PUFAs) from *Chlorella* spp. provides a sustainable route for obtaining omega-3 and omega-6 lipids for food, nutraceutical, and aquafeed applications. This protocol outlines a standardized laboratory-scale workflow for producing PUFA-rich oils from microalgal biomass. The chapter details biomass preparation, cell-wall disruption, solvent-based lipid extraction, FAME/FAEE derivatization, and GC–FID quantification of fatty-acid profiles. An optional urea-complexation step is included to enhance the proportion of unsaturated fatty acids. Emphasis is placed on temperature control, antioxidant protection, solvent handling, and nitrogen blanketing to preserve lipid integrity throughout processing. Quality control measures such as oxidation indices and analytical validation steps are incorporated to ensure reproducibility and product stability. The protocol also provides troubleshooting guidance for improving lipid yield, PUFA enrichment, and chromatographic resolution.

Key words *Chlorella*, PUFA extraction, Omega-3 fatty acids, Omega-6 fatty acids, Lipid extraction, Microalgae bioprocessing, ALA, GLA, SDA, Solvent extraction

1 Introduction

Polyunsaturated fatty acids (PUFAs), particularly omega-3 and omega-6 species, play critical roles in maintaining membrane fluidity, regulating inflammatory pathways, and supporting cardiovascular and neural health. Microalgae, especially species of the genus *Chlorella*, represent a sustainable and vegetarian source of PUFAs for food, nutraceutical, and aquafeed applications. Among these, linoleic acid (LA; 18:2 n-6) and alpha-linolenic acid (ALA; 18:3 n-3) are the dominant fatty acids in *Chlorella* biomass, with their relative abundance varying according to strain, cultivation conditions, and nutrient status. This chapter outlines a standardized laboratory-scale procedure for extracting and enriching omega-3 and omega-6 PUFAs from *Chlorella* spp. Key steps include

Dhanasekaran Dharumadurai (ed.), *Chlorella Genomics, Cultivation, and Bioenterprise: Green Economy through Algal Innovations*, Springer Protocols Handbooks, https://doi.org/10.1007/978-1-0716-5276-3_32,
© The Author(s), under exclusive license to Springer Science+Business Media, LLC, part of Springer Nature 2026

controlled cell wall disruption, solvent-based lipid extraction, fatty acid methyl ester (FAME) preparation, and gas chromatographic quantification. An optional PUFA enrichment step using urea complexation is included to increase the relative proportion of unsaturated fatty acids. Emphasis is placed on optimizing parameters such as extraction temperature, solvent choice, antioxidant use, and nitrogen blanketing to preserve lipid integrity. The protocol also includes critical quality control checkpoints such as oxidation indices (peroxide value, p-anisidine value) and provides reproducible methods for downstream compositional analysis.

2 Materials

Prepare all reagents using analytical-grade chemicals. Ensure all tools, containers, and surfaces are lab-grade, clean, and free from contamination. Handle all solvents and gases in well-ventilated areas or fume hoods, and store temperature-sensitive materials appropriately.

Reagents and Solvents

1. **Antioxidants:** Butylated hydroxytoluene (BHT) or butylated hydroxyanisole (BHA); prepare 1% (w/v) stock solution and add during extraction and handling.
2. **Extraction solvents** (select based on intended application):
   (a) Lab-grade ethanol (95–100%).
   (b) *n*-Hexane or isopropanol: *n*-hexane mixture (3:2).
   (c) Chloroform: methanol (2:1)—for analytical extractions (Folch/Bligh–Dyer methods).
   (d) Methyl tert-butyl ether (MTBE) —optional for lipidomics.
3. **Urea complexation:** Urea (analytical grade), ethanol or green solvent alternative, deionized water.
4. **FAME/FAEE preparation:** Methanol or ethanol; methanolic HCl or BF_3-methanol reagent; anhydrous sodium sulfate (Na_2SO_4); hexane or iso-octane for extraction.
5. **Oxidation quality control:** Peroxide value (PV) and p-anisidine value (p-AnV) assay kits or reagents.
6. **Inert gases:** High-purity nitrogen or argon for solvent blanketing and lipid storage.

Standards and Calibration

7. Certified FAME standard mix (e.g., 37-component mix) for peak identification and response factor calibration.

8. Internal standard: Methyl nonadecanoate (C19:0) for quantitative GC analysis.
9. Optional: Dimethyl disulfide (DMDS) for double-bond position confirmation by GC–MS.

Equipment

10. **Cell disruption:** Bead mill or high-pressure homogenizer (HPH); optional ultrasonic probe or bath.
11. **Separation:** Refrigerated centrifuge ($\geq$10,000× *g*); vacuum filtration unit with compatible glassware.
12. **Solvent removal:** Rotary evaporator with adjustable vacuum control and cold-trap condenser.
13. **Chromatography:** Gas chromatograph with flame ionization detector (GC–FID) and high-polarity capillary column (e.g., CP-Sil 88, DB-23, or SP-2560); GC–MS optional for structural confirmation.
14. **Supercritical fluid extraction (optional):** SFE-CO_2 system equipped with ethanol co-solvent pump.
15. **Cold processing and storage:** −20 °C freezer; 0–5 °C cold room or refrigerator for urea crystallization steps.
16. **General lab tools:** Analytical balance (precision ±0.1 mg), amber vials, vortex mixer, glass water bath ($\leq$70 °C), calibrated thermometer.

Consumables

17. Polypropylene centrifuge tubes (15 mL and 50 mL).
18. PTFE syringe filters (0.45 μm), Whatman filter papers.
19. Amber glass bottles and sample vials (airtight).
20. Solvent-resistant labels and permanent markers.

Safety and Handling

21. **Personal protective equipment (PPE):** Lab coat, nitrile gloves, and chemical splash goggles.
22. Use a certified fume hood for handling volatile or halogenated solvents.
23. Exercise caution when working with **p-anisidine** and **BF_3-methanol** due to corrosive and toxic properties; consult SDS and institutional safety protocols.
24. Segregate and label **halogenated** and **nonhalogenated** solvent waste; dispose of according to regulatory guidelines.

3 Methods

All procedures should be conducted at room temperature unless otherwise specified. Ensure solvent contact steps are performed under a fume hood, and antioxidant protection (e.g., BHT) is applied throughout lipid handling.

3.1 Biomass Preparation

1. Dewater and dry *Chlorella* biomass using a freeze dryer, or a low-temperature oven set at ≤45–50 °C.
2. Grind the dried biomass into a fine powder (≤500 μm) using a laboratory mill.
3. Store powder at −20 °C in amber, nitrogen-flushed containers to minimize oxidation.

Validation Step: Final moisture content ≤10%; homogeneous particle size enhances extraction efficiency.

3.2 Cell Wall Disruption (Choose One Method Based on Available Equipment)

1. **Bead Milling**
   (a) Prepare a 10–20% (w/v) biomass slurry in cold distilled water or 95% ethanol.
   (b) Add zirconia beads (0.5–1.0 mm).
   (c) Mill for 5–10 cycles (1 min on/1 min off), maintaining slurry temperature <30 °C using an ice bath.
   (d) Remove beads and proceed to extraction.
2. **High-Pressure Homogenization (HPH)**
   (a) Prepare slurry at 10–15% (w/v).
   (b) Pass through the homogenizer at 300–900 bar for 1–5 cycles. Cool between passes.
   (c) Collect homogenate for extraction.
3. **Ultrasonication**
   (a) Treat biomass using probe or bath sonication (20–40 kHz) in pulse mode (e.g., 10 s on/10 s off for 5–10 min).
   (b) Maintain the sample below 30 °C using an ice bath.
4. **Enzymatic Treatment (Food-safe)**
   (a) Suspend biomass in buffer and add cell wall degrading enzymes (e.g., lysozyme, pectinase, chitinase).
   (b) Incubate for 30–60 min at 30–40 °C. Proceed directly to lipid extraction.

Validation Step: Use microscopy or turbidity measurement to confirm ≥80% disruption.

3.3 Lipid Extraction (Select a Route Based on Application)

Step 1—Food-Grade Ethanol Extraction (Preferred for Edible PUFA Oil)

1. Mix disrupted biomass with 95% ethanol (1:5 to 1:10 w/v); add antioxidant (e.g., 1% BHT, a few drops per 100 mL).
2. Stir for 1–2 h at 40–50 °C under nitrogen.

3. Centrifuge and collect supernatant. Repeat extraction once; combine both.
4. (Optional) Wash with brine to separate emulsified layers.
5. Evaporate solvent under reduced pressure ($\leq$40 °C).

Step 2—Bligh-Dyer/Folch Method (For Analytical Comparison Only)

1. Homogenize biomass in chloroform: methanol (2:1).
2. Add water to induce phase separation.
3. Recover the lower (organic) phase, wash, dry over Na_2SO_4, and evaporate solvent.

Step 3—Supercritical CO_2 Extraction (SFE)

1. Load dried biomass into the extractor.
2. Run at 110–250 bar and 40–60 °C; co-solvent (ethanol, 5–15%) improves PUFA yield.
3. Collect and pool lipid-rich fractions.

Validation Step: Record crude lipid yield (% DW), visual clarity, and color.

3.4 Crude Oil Work-Up and Storage

1. Dry lipid extract over anhydrous sodium sulphate (Na_2SO_4) and filter.
2. Pass through a short silica bed to reduce pigments. (Optional)
3. Flush with nitrogen, add antioxidant (e.g., tocopherols, 200–500 ppm).
4. Store in amber glass vials at −20 °C.

3.5 FAME (or FAEE) Preparation for GC Analysis

1. Aliquot crude oil or enriched fraction.
2. Trans-esterify using methanolic HCl (for FAME) or ethanolic HCl (for FAEE).
3. Extract esters into hexane or iso-octane; dry with Na_2SO_4.
4. Add internal standard (e.g., methyl nonadecanoate, C19:0).
5. Store at −20 °C in amber vials under nitrogen until GC analysis.

3.6 GC–FID Analysis of Fatty Acid Profile

1. Use a high-polarity capillary column (e.g., CP-Sil 88, DB-23, or SP-2560).
2. Apply an ISO 12966-4 compliant temperature program for optimal PUFA separation.
3. Identify and quantify using a certified FAME mix.
4. Report results as wt.% of each fatty acid and as grouped totals: Σn-3, Σn-6, PUFA, MUFA, SFA.

Validation Step: Replicate RSD ≤2% for major FAME peaks ensures analytical precision.

3.7 PUFA Enrichment via Urea Complexation

Urea selectively forms adducts with saturated and monounsaturated FAs/FAEEs; PUFAs remain unbound and are recovered in the filtrate.

Free Fatty Acids (FFA):

1. Hydrolyze triglycerides to FFAs using alkaline saponification.
2. Dissolve urea in warm ethanol (60–70 °C).
3. Combine FFAs and urea at a 3:1 (w/w) ratio; optimize 2:1–4:1.
4. Crystallize at 0–4 °C for 12–24 h.
5. Filter cold; collect filtrate as PUFA-rich fraction.
6. Evaporate ethanol, dry lipid, flush with nitrogen, and store at −20 °C.

Fatty Acid Ethyl Esters (FAEE):

Same as above but use FAEEs instead of FFAs; convert back to TAGs if needed post-enrichment.

Validation Step: Confirm increased PUFA% and reduced SFA/-MUFA by GC–FID.

3.8 Oxidation Control and Quality Check

1. Maintain low temperature, limited light exposure, and nitrogen protection throughout.
2. Measure oxidative indices:
   (a) **Peroxide Value (PV)**
   (b) **p-Anisidine Value (p-AnV)**
   (c) **TOTOX Value** = 2 × PV + p-AnV
3. Accept/reject based on defined quality thresholds (e.g., Codex/GOED standards).

3.9 Calculations (Report All Values as Mean ± SD)

1. **Lipid yield (% DW)** = (mass of crude lipid/dry biomass) × 100
2. **PUFA (% of total FAME)** = 100 × (Σ PUFA peak areas/Σ all FAME peak areas)
3. **n-3/n-6 ratio** = (Σ n-3 FAME)/(Σ n-6 FAME)
4. **Enrichment factor** = PUFA% (after enrichment)/PUFA% (before enrichment)

Include internal standard-corrected mass recovery per ISO 12966 guidance.

3.10 Statistical Analysis

1. Conduct all extractions in biological triplicate.
2. Analyze variance using one-way ANOVA; apply Tukey's HSD post hoc test ($\alpha = 0.05$).
3. Use linear regression or polynomial models to evaluate the impact of urea: FA ratio and crystallization temperature on PUFA enrichment outcomes.

Implementation of this PUFA extraction protocol consistently produced high-quality, oxidation-stable, PUFA-rich lipid fractions from *Chlorella* biomass. Dried biomass retained a moisture content between 4–10%, enabling efficient solvent penetration during extraction. Cell disruption methods such as ultrasonication and bead milling yielded 75–90% disruption efficiency, resulting in 15–25% total lipid recovery from the dry biomass (Table 1). The crude lipid fraction showed a PUFA contribution of 30–45% of total fatty acids, with clear GC-FID/GC-MS peaks corresponding to ALA (C18:3 n-3), SDA (C18:4 n-3), LA (C18:2 n-6), and GLA (C18:3 n-6). Use of inert conditions (N_2 flushing + BHT) kept oxidation minimal, confirmed by low MDA values and absence of rancid odor. Purification via urea complexation and silica gel chromatography significantly enriched the PUFA fraction to 60–85%, depending on the method. The final PUFA extract appeared light yellow to pale green, with low viscosity and minimal pigment carryover, indicative of successful chlorophyll removal (*See* **Notes 1–6**).

4 Notes

1. Freeze-drying is the optimal pre-processing method for PUFA preservation, as it retains pigments, prevents thermal oxidation, and enhances subsequent extraction efficiency [3, 5, 10].
2. Mechanical cell disruption (bead milling/ultrasonication) is superior for lipid release from the thick *Chlorella* cell wall, ensuring higher lipid and PUFA yields than enzymatic treatments alone (Table 2) [5, 9, 16].
3. Solvent polarity determines PUFA recovery, with the Bligh and Dyer method yielding the cleanest extracts and sharpest chromatographic profiles [1, 2, 5].
4. Urea complexation and silica chromatography demonstrate strong PUFA selectivity, enabling removal of saturated fats and pigment impurities [10, 16].
5. Antioxidant and oxygen-free processing conditions are essential to preserve PUFA integrity during extraction and purification [6, 10, 17].

Table 1

Experimental possible outcomes obtained during the preparation of omega-3 and omega-6 polyunsaturated fatty acids (PUFAs) from *Chlorella* biomass

Parameter evaluated	Expected results	Scientific inference/interpretation
Biomass moisture after drying	4–8% (freeze-drying), 6–10% (tray drying)	Low moisture prevents lipid oxidation and increases extraction efficiency. Freeze-drying gives the highest PUFA preservation
Cell disruption efficiency	Ultrasonication: 75–85%; Bead milling: 80–90%; Enzymatic hydrolysis: 60–70%	Mechanical disruption gives higher lipid release. Enzymatic hydrolysis is milder but preserves heat-sensitive PUFAs
Total lipid yield	15–25% of dry biomass	Yield aligns with typical *Chlorella* lipid content. Efficient cell disruption increases total lipid recovery
PUFA concentration in crude lipids	30–45% of total fatty acids	Indicates *Chlorella* biomass is a good source of ALA, LA, and GLA
Omega-3 content (ALA + SDA)	ALA: 20–35%; SDA: 1–5%	High ALA indicates strong nutritional potential for aquafeeds and nutraceuticals
Omega-6 content (LA + GLA)	LA: 10–25%; GLA: 1–7%	Balanced omega-6 profile supports anti-inflammatory benefits and energy metabolism in fish
Purification method efficiency	Urea complexation: 60–75% PUFA enrichment; cold crystallization: 40–55%; silica column: 70–85%	Silica column gives highest purity; urea complexation gives best yield-to-purity balance. Cold crystallization is economical but less selective
PUFA oxidation index (MDA)	Very low when BHT + N_2 flushing used	Antioxidant protection is essential; oxidation increases sharply without inert handling
GC-FID/GC-MS chromatogram quality	Clear peaks for C18:2, C18:3, C18:4	Confirms successful extraction, conversion to FAMEs, and PUFA identification
Final PUFA-rich oil appearance	Light yellow to pale green; low viscosity	Color and clarity indicate minimal pigment contamination and successful chlorophyll removal

The table integrates biomass quality parameters, extraction yields, PUFA profiles, purification efficiencies, and oxidative stability indicators, along with scientific interpretations. These observations collectively validate the effectiveness of the optimized extraction workflow and highlight the influence of drying, cell disruption, solvent choice, and purification method on the final PUFA-rich lipid quality

Table 2
Typical PUFA Profile in *Chlorella* Species

PUFA	Common name	Typical range (%) of total fatty acids
α-Linolenic acid (ALA, C18:3 n-3)	Omega-3	20–35%
Stearidonic acid (SDA, C18:4 n-3)	Omega-3	1–5%
Linoleic acid (LA, C18:2 n-6)	Omega-6	10–25%
γ-Linolenic acid (GLA, C18:3 n-6)	Omega-6	1–7%
Oleic acid (C18:1)	MUFA	10–20%

6. GC-FID/GC-MS verification confirms protocol reliability, demonstrating reproducible separation and quantification of omega-3 and omega-6 fatty acid profiles [5, 10, 16].

Overall, the protocol effectively produces stable, enriched PUFA fractions suitable for nutraceutical, aquafeed, and functional food applications.

When the protocol is executed correctly, the following outcomes should be observed:

Biomass and Preprocessing

1. Dry powder moisture: 4–10% [3, 4]
2. No browning, no rancid smell [6, 10]
3. Fine particle size (<500 μm) for efficient extraction [5, 16]

Cell Disruption

1. Ultrasonication: 75–85% efficiency [9, 16]
2. Bead milling: 80–90% efficiency [5, 16]
3. Enzyme hydrolysis: 60–70% efficiency [4, 10]
4. Biomass becomes visibly lighter and more fragmented [9, 17]

Lipid Extraction

1. Total lipid yield: 15–25% of dry biomass [5, 17]
2. Extract color: greenish–yellow transitioning to pale yellow after purification [1, 10]
3. Low foam, minimal emulsion formation (Fig. 1) [1, 2]

PUFA Composition (Typical for *Chlorella* spp.)

1. ALA (C18:3 n-3): 20–35% [10, 15]
2. SDA (C18:4 n-3): 1–5% [10, 17]
3. LA (C18:2 n-6): 10–25% [3, 10]
4. GLA (C18:3 n-6): 1–7% [10, 15]

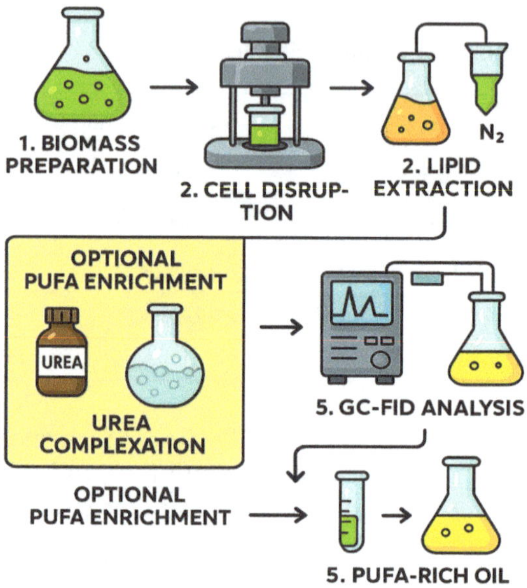

Fig. 1 Workflow for extraction and purification of polyunsaturated fatty acid (PUFA)-Rich Lipids from *Chlorella* spp. commencing with biomass harvesting, where wet microalgal paste is collected and concentrated, followed by low-temperature drying to preserve thermolabile lipids. The dried biomass undergoes cell wall disruption through mechanical, ultrasonic, or high-pressure methods to release intracellular oils. Lipids are then recovered using solvent extraction (chloroform–methanol, hexane–isopropanol, or ethanol) under nitrogen protection to minimize oxidation. The extracted crude oil proceeds to an optional PUFA enrichment module, which may involve urea complexation to remove saturated fatty acids, cold crystallization to enhance unsaturated fractions, or silica-gel chromatography to purify PUFA-rich components. The final lipid extract is converted to FAME/FAEE derivatives and analyzed using GC–FID or GC–MS, providing quantitative profiles of omega-3 and omega-6 fatty acids. This integrated workflow ensures efficient recovery, stabilization, and analytical validation of PUFA-rich algal lipids for nutraceutical, food, and aquafeed applications

Purification Outcomes

1. Urea complexation: 60–75% PUFA enrichment [10, 16]
2. Silica column: 70–85% PUFA enrichment [10, 16]
3. Low-temperature crystallization: 40–55% PUFA enrichment [10]

Oxidative Stability

1. Very low MDA levels [6, 17]
2. No darkening or rancidity [3, 10]
3. Stable PUFA composition over short-term storage ($\leq$4 °C, dark, N_2) [6, 16]

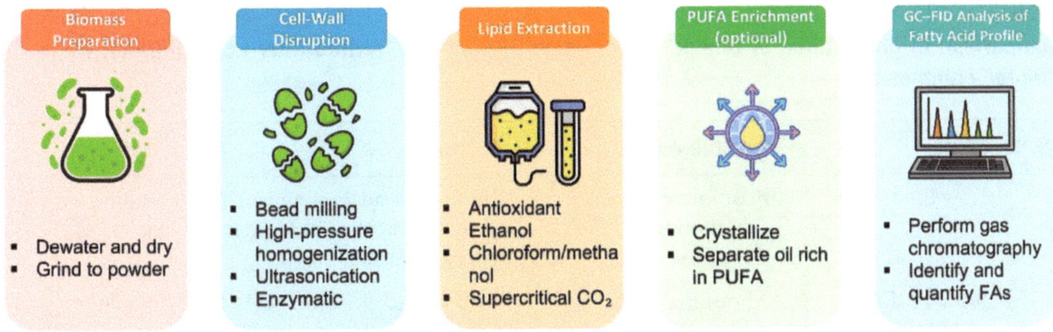

Fig. 2 Schematic representation of the PUFA extraction and analysis process from *Chlorella* biomass

GC-FID/GC-MS Analysis

1. Clear, well-resolved peaks for C18 PUFAs [5, 10]
2. Low noise, minimal chlorophyll interference [1, 5]
3. FAME conversion efficiency >95% [5, 16]

Final Product

1. PUFA-rich oil suitable for: Aquafeed enrichment; Nutraceutical formulations; Biochemical assays; Functional food development (Fig. 2).

The Following Points to Considered During the Experiment:

1. Maintain low-temperature and low-light conditions during all lipid handling steps to minimize PUFA oxidation.
2. Add antioxidants (e.g., BHT, BHA, or tocopherols) to solvents before extraction to prevent auto-oxidation of sensitive unsaturated fatty acids.
3. Ensure complete cell wall disruption; insufficient disruption significantly reduces lipid yield and PUFA recovery.
4. Keep biomass moisture below 10% before extraction, as excess water interferes with solvent penetration and phase separation.
5. During urea complexation, strict temperature control (0–4 °C) is essential to achieve selective crystallization and high PUFA enrichment.
6. Use only amber or opaque glassware for lipid storage to prevent pigment and PUFA degradation caused by light exposure.
7. Avoid overheating during solvent evaporation (≤40 °C), as prolonged heating accelerates oxidation and loss of polyunsaturated lipids.
8. Regularly flush storage containers with nitrogen or argon to maintain an oxygen-free environment for PUFA stability.

Table 3
Comparison of preparation of omega-3 and omega-6 polyunsaturated fatty acids (PUFAs) from *Chlorella* biomass

S.No	Issue	Possible cause	Remedy
1.	Low lipid yield	Incomplete cell disruption	Increase bead-milling cycles or combine with ultrasonication; consider enzyme pre-treatment
2.	Poor PUFA enrichment	Suboptimal urea:FA ratio or crystallization conditions	Adjust urea:FA ratio (2:1–4:1), crystallize longer/colder, or use FAEEs instead of FFAs
3.	FAME co-elution or bad GC resolution	Unsuitable GC column or temperature program	Use highly polar column (CP-Sil 88/DB-23) with ISO 12966-4 compliant GC ramp
4.	High PV/p-AnV after processing	Lipid oxidation during heat/air exposure	Add antioxidants early, limit processing time, work under nitrogen, and keep ≤50 °C
5.	Green pigment carry-over	Chlorophyll co-extraction	Use brief food-grade adsorbent polish before enrichment; work under dim light
6.	PUFA levels lower than expected	Biological strain limitation (C18 dominance)	Acknowledge strain limits; focus on ALA/LA products or consider EPA-rich taxa like *Nannochloropsis*

9. For GC-FID analysis, ensure column conditioning and calibration using certified FAME mixes to avoid peak shifts or co-elution errors (Table 3).
10. Dispose of halogenated and nonhalogenated solvent waste separately to comply with laboratory safety and environmental regulations.

Safety and Regulatory Considerations

1. **Solvents:** Use chloroform:methanol only in fume hoods with proper halogenated waste disposal. Prefer ethanol or supercritical CO_2 for food-grade extractions.
2. **Oxidation Control:** Add BHT/BHA during extraction and storage. Follow Codex or GOED limits for antioxidant use and oxidation indices (PV, p-AnV, TOTOX).
3. **Urea Complexation:** Use food-grade solvents. Apply green protocols to reduce ethyl carbamate risk during crystallization.

References

1. Bligh EG, Dyer WJ (1959) A rapid method of total lipid extraction and purification. Can J Biochem Physiol 37(8):911–917
2. Folch J, Lees M, Stanley GS (1957) A simple method for the isolation and purification of total lipides from animal tissues. J Biol Chem 226(1):497–509
3. Becker EW (2013) Microalgae for human and animal nutrition. In: Handbook of microalgal

culture: applied phycology and biotechnology, pp 461–503
4. de Andrade CJ, de Andrade LM (2017) An overview on the application of genus Chlorella in biotechnological processes. J Adv Res Biotechnol 2:1–9
5. Ryckebosch E, Muylaert K, Foubert I (2012) Optimization of an analytical procedure for extraction of lipids from microalgae. J Am Oil Chem Soc 89(2):189–198
6. Lam MK, Lee KT (2012) Microalgae biofuels: a critical review of issues, problems and the way forward. Biotechnol Adv 30(3):673–690
7. Radhakrishnan S, Seenivasan C, Muralisankar T (2015) Effect of dietary replacement of fishmeal with Chlorella vulgaris on growth performance, energy utilization and digestive enzymes in Macrobrachium rosenbergii postlarvae. Int J Fish Aquac 7(5):62–70
8. Borowitzka MA, Moheimani NR (2013) Algae–culture and applications 7. Algae for energy and feed: A, 79
9. Halim R, Danquah MK, Webley PA (2012) Extraction of oil from microalgae for biodiesel production: a review. Biotechnol Adv 30(3): 709–732
10. Medina AR, Grima EM, Giménez AG, González MI (1998) Downstream processing of algal polyunsaturated fatty acids. Biotechnol Adv 16(3):517–580
11. Ma M, Hu Q (2024) Microalgae as feed sources and feed additives for sustainable aquaculture: prospects and challenges. Rev Aquac 16(2):818–835
12. McCleary BV (2023) Measurement of dietary fiber: which AOAC official method of analysis SM to use. J AOAC Int 106(4):917–930
13. NRC (1993) Nutrient requirements of fish. National Academy Press, Washington, DC
14. Ozcelik D, Espinosa R, Shetty R, Suwal S, Røder HL, Jacobsen C, Poojary MM. Biotransformation of lactose free acid whey into high value microalgal biomass. Available at SSRN 5414321
15. Ng DH, Ng YK, Shen H, Lee YK (2015) Microalgal biotechnology: the way forward. In: Handbook of marine microalgae. Academic, pp 69–80
16. Lee SY, Khoiroh I, Vo DVN, Senthil Kumar P, Show PL (2021) Techniques of lipid extraction from microalgae for biofuel production: a review. Environ Chem Lett 19(1):231–251
17. Hu Q, Sommerfeld M, Jarvis E, Ghirardi M, Posewitz M, Seibert M, Darzins A (2008) Microalgal triacylglycerols as feedstocks for biofuel production: perspectives and advances. Plant J 54(4):621–639
18. Babu SS, Gondi R, Vincent GS, JohnSamuel GC, Jeyakumar RB (2022) Microalgae biomass and lipids as feedstock for biofuels: sustainable biotechnology strategies. Sustainability 14(22):15070
19. Araújo RDS (2017) Expressão gênica em frangos de corte submetidos a diferentes níveis de alimentos alternativos
20. Jacob-Lopes E, Queiroz MI, Maroneze MM, Zepka LQ (eds) (2023) Handbook of food and feed from microalgae: production, application, regulation, and sustainability. Elsevier

Chapter 33

Preparation and Quality Assessment of *Chlorella* Green Smoothie

Shrika Kumaresan and Dhanasekaran Dharumadurai

Abstract

Chlorella vulgaris has become a versatile ingredient in functional beverages, such as dairy, plant-based, and alcoholic drinks. This chapter describes a method for preparing drinks based on *Chlorella*, with a focus on creating smoothies and other nondairy functional beverages for health-conscious vegan consumers, using *C. vulgaris* powder as a source of protein, pigments, and antioxidants. The formulation describes the dispersion of *Chlorella* in aqueous or plant-based matrices with analytical evaluation of physicochemical properties, color analysis, microbial load, and sensory attributes.

Key words *Chlorella vulgaris*, Functional beverages, Nondairy formulations, Physicochemical analysis, Smoothie development

1 Introduction

Functional beverages available in the market today generally fall into three major categories: dairy-based drinks, fruit or vegetable beverages, and sports or energy drinks [1]. The beverage sector has become the most active area of research for incorporating microalgal-derived bioactive compounds. Particularly, plant-based and nondairy milk beverages are one of the fastest-growing sectors worldwide. The demand for these products is growing, as 79 million people around the world are adopting a vegan lifestyle [2]. In this regard, microalgae have gained special attention, because they can enhance the natural colors of food products, such as green or orange tones, making them more attractive to vegan consumers. In addition to their coloring and functional benefits, microalgae-derived proteins have great potential to meet global protein needs. They have several benefits over traditional protein sources, especially regarding sustainability. For instance, producing 1 kg of microalgal protein needs less than 2.5 m^2 of land, which is much

Dhanasekaran Dharumadurai (ed.), *Chlorella Genomics, Cultivation, and Bioenterprise: Green Economy through Algal Innovations*, Springer Protocols Handbooks, https://doi.org/10.1007/978-1-0716-5276-3_33,
© The Author(s), under exclusive license to Springer Science+Business Media, LLC, part of Springer Nature 2026

more efficient than animal protein production such as pork (47–64 m^2/kg protein), chicken (42–52 m^2/kg protein), and beef (144–258 m^2/kg protein) [3].

Among several microalgae, *Chlorella vulgaris* stands out as a valuable microalgal ingredient for these products, as it naturally contains a wide range of beneficial compounds, including canthaxanthin, astaxanthin, bioactive peptides, and oleic acid [4]. It contains high amount of essential fatty acid precursors, particularly α-linolenic acid (ALA) from the omega-3 family(ω3) and linoleic acid (LA) from the omega-6 family (ω6). These are essential fatty acids to the human body, as they cannot synthesize on their own and must obtain them through diet. A major advantage of *Chlorella vulgaris* is high protein content, around 50–60% of its dry biomass, with a well-balanced essential amino acid composition [5]. As a result, *Chlorella*-based beverages naturally serve as concentrated sources of plant-derived protein source, contributing to improved satiety, muscle maintenance, and overall well-being.

However, several technical and sensory challenges are need to be addressed while formulating *Chlorella*-based drinks. The grassy taste and marine-like odor of *C. vulgaris* may limit the consumer's acceptance if not properly masked or balanced with other fruit, vegetable, or artificial flavors. In addition to this, the cell structure and particle size of the biomass influence dispersion stability, mouthfeel, viscosity, and sedimentation, which is important for smoothie quality and shelf life [6]. This chapter provides a stepwise protocol for the preparation of *Chlorella*-based drinks, focusing on nondairy, plant-origin matrices with options for several vegan juice and smoothie variants. The protocol covers raw material selection and preparation of green smoothie using *C. vulgaris* powder. Methods for assessing pH, titratable acidity, viscosity, color, soluble solids content, microbial counts, and sensory attributes are included to enable comprehensive evaluation of beverage quality and functionality.

2 Materials

2.1 Raw Material Preparation

1. White seedless grapes.
2. Broccoli.
3. Cucumbers.
4. *Chlorella vulgaris* powder (food grade).
5. 0.1% NaClO.
6. Water bath or blanching tank (stainless steel, 10 L capacity, controlled temperature 85–100 °C).
7. Digital thermometer (0–150 °C, 0.1 °C precision).
8. Wire mesh baskets (perforated stainless steel, 20 × 20 cm).

9. Ice water bath (0–4 °C).
10. Timer (digital, 0–60 min).
11. Perforated spoon/scoop (stainless steel).

2.2 Green Smoothie Preparation

1. Blender.

2.3 Thermal Treatment

1. High-pressure homogenizer (Eg; GEA Ariete Series, Raffaello Series).

2.4 Aseptic Packaging and Storage

1. Sterile glass jars.

2.5 Physicochemical Analysis

2.5.1 pH Measurement

1. pH meter.
2. pH standard buffers (pH–4.0, pH–7.0).

2.5.2 Titratable Acidity

1. 0.1 N NaOH (freshly standardized daily).
2. Phenolphthalein indicator (0.1% w/v in 95% ethanol).
3. 50 mL burette (0.01 mL graduation).
4. pH meter.

2.5.3 Soluble Solids Content

1. Handheld refractometer (0–32°Brix, Atago PAL-1).

2.6 Rheology

1. Stress- controlled rheometer (ARG2, TA Instrument, New Castle, DE, USA).
2. Serrated plate–plate geometry (20 mm diameter, 2 mm Gap).
3. Peltier temperature control system (20 °C0).
4. 5-mL syringes (sample loading).
5. Light paraffin oil (edge covering).
6. TA Orchestrator software (v3.2.0).

2.7 Color Analysis

1. Colorimeter (Minolta CM-2600d or Konica Minolta CR-400).
2. White calibration plate.
3. Transparent glass tubes (20 mm diameter × 50 mm height).
4. Black matte background cloth.

2.8 Microbiological Properties

1. Sterile Petri dishes (90 mm).
2. Laminar flow hood.

3. Incubator.
4. L-rod.
5. Peptone water (0.1%).
6. Plate count agar.
7. Dichloran-Rose Bengal-Chloramphenicol Agar (DRBC).
8. Eosin Methylene Blue agar.
9. Agar Listeria Ottavani and Agosti (ALOA Agar).
10. Xylose Lysine Deoxycholate agar.

2.9 Sensory Properties

1. Three-digit random coding labels.
2. Mineral water (palate cleanser).
3. Unsalted crackers (neutralizer).
4. Sensory evaluation scorecards.
5. Ten Individual taste booths.

2.10 General Laboratory Consumables

1. Analytical balance (±0.0001 g).
2. 50 mL and 100 mL volumetric flasks.
3. Pipettes (10 mL, 5 mL).
4. Micropipettes (1 mL, 200 μL) and tips.
5. Beakers.
6. 50 mL, 100 mL Erlenmeyer flasks.
7. Tissue papers.

3 Methods

3.1 Raw Material Preparation

1. Select fresh, undamaged white seedless grapes, broccoli, and cucumbers.
2. Sanitized all fresh produce with 0.1% NaClO (Sodium hypochlorite solution) for 2 min and then rinse with tap water for 2 min.
3. Prepare ingredients:
   (a) Grapes: Destem and halve longitudinally (no peeling required).
   (b) Broccoli: Cut into 2 cm florets including tender stems.
   (c) Cucumber: Wash thoroughly and cut into 2 cm thick slices with skin.
4. Blanch the Broccoli and cucumber in boiling water (95 °C) for 1–2 min. No blanching for grapes.

5. Transfer blanched vegetables to ice water bath (0–4 °C) for 2 min.
6. Drain excess water using clean kitchen towels and pat dry [7].

3.2 Green Smoothie Preparation

1. Weigh ingredients exactly according to the table below (±0.01 g precision).

Ingredient	Control	*Chlorella* Smoothie
White seedless grapes (*Vitis vinifera* cv. Sugraone)	57.8 g	56.5 g
Broccoli	15.8 g	15.5 g
Cucumber (*Cucumis sativus*)	26.4 g	25.8 g
Chlorella vulgaris (dry powder)	0 g	2.2 g

2. Combine all weighed ingredients (except *Chlorella* powder) in blender.
3. Blend at 10,000 rpm for 3 min until homogeneous mixture forms with particle size <0.5 mm (*see* **Note 1**).
4. Add *Chlorella vulgaris* powder during final 30 s of blending. No straining required (to preserve the fiber content) [7].

3.3 Thermal Treatment

1. Fill the smoothie in 50 mL Falcon tubes in 85–90 °C water bath for 6 min.
2. Immediately cool in Ice water bath (0–4 °C) until <10 °C for 2 min.
3. Homogenize the juice mixture using High-pressure homogenization at 150 bar, 2 passes for Industrial scale production and Ultra-Turrax T25 at 10,000 rpm for 2 min for laboratory scale [8].

3.4 Aseptic Packaging and Storage

1. Fill sterile glass jars (100-mL portions) under laminar flow hood.
2. Seal hermetically.
3. Store at 5 ± 1 °C in dark conditions.
4. Analyze at 0, 7, 14, 21, 24 days storage [7].

3.5 Physicochemical Analysis

3.5.1 pH Measurement

1. Calibrate pH meter with pH 4.0, 7.0 standards at room temperature.
2. Immerse electrode in 20 mL sample, stir gently (100 rpm)
3. Record stable reading after 30 s [9].

3.5.2 Titratable Acidity

1. Mix the *Chlorella* drink thoroughly.
2. Pipette 10 mL of sample into a clean conical flask. Dilute with 10–20 mL distilled water (preferably 1:1) to facilitate accurate titration.
3. Add 3 drops of phenolphthalein indicator (0.1% w/v in 95% ethanol).
4. Titrate with 0.1 N NaOH under continuous swirling.
5. Continue titration until a persistent pale-pink color appears (30 s) or monitor the pH until it reaches 8.2–8.3 (*see* **Note 2**).
6. Note the volume (mL) of NaOH used.
7. Repeat with 20 mL distilled water only for control blank [10].
8. Calculation: Express acidity as % citric acid (or other organic acid based on formulation):

$$\text{TA (\%citric acid)} = \frac{V_{\text{NaOH}} \times N_{\text{NaOH}} \times 0.064 \times 100}{\text{Sample (mL)}}$$

(0.064 = milliequivalent factor for citric acid)

3.5.3 Soluble Solids Content (SSC) Determination

1. Calibrate the digital hand-held refractometer (Atago N1, Tokyo, Japan) with distilled water at 25 °C to ensure accuracy.
2. Place 2–3 drops of well-homogenized *Chlorella* smoothie sample on the prism surface.
3. Close the cover, wait for stable reading (10–15 s), and record SSC expressed as °Brix at 25 °C.
4. Wipe the prism clean with distilled water and dry between readings. Perform triplicate measurements per sample [11].

3.6 Color Analysis

3.6.1 Equipment and Calibration

1. Use Konica Minolta CM-2600d colorimeter (Japan) with light source C, 2° observer, and 8-mm viewing aperture.
2. Calibrate with white reference plate (CM-A145) at 23 °C prior to measurements for absolute colorimetric accuracy.

3.6.2 Sample Preparation and Measurement

1. Fill a special glass tube (transparent, cylindrical) with well-homogenized *Chlorella* smoothie sample, ensuring no air bubbles.
2. Mount the tube securely on the colorimeter's sample holder device.
3. Measure at three equidistant points along the tube by rotating it 120° between readings.

4. Record average L* (lightness), a* (red-green), b* (yellow-blue) values in CIE L*a*b* color space automatically [11] (*see* **Note 4**).
5. Total color difference (ΔE) Calculation: Calculate ΔE throughout storage relative to day 0 value using:

$$\Delta E = \sqrt{\left(L_t^* - L_0^*\right)^2 + \left(a_t^* - a_0^*\right)^2 + \left(b_t^* - b_0^*\right)^2}$$

3.7 Rheology

3.7.1 Instrument Preparation

1. Use a stress-controlled rheometer (ARG2, TA Instruments, New Castle, DE, USA) equipped with a serrated plate–plate geometry.
2. Clean and dry the serrated parallel-plate geometry.
3. Install the 20 mm upper plate and set a 2-mm gap and set temperature control to 25 °C.
4. Fill the solvent trap with distilled water to prevent evaporation (Table 1).

3.7.2 Sample Loading

1. Load 5 mL homogenized *Chlorella* smoothie via syringe to the lower plate and lower the upper plate to the 2-mm gap.
2. Remove excess sample at the rim and allow the sample to equilibrate for 1 min at 25 °C under the solvent trap. Avoid air bubbles (*see* **Note 3**).

3.7.3 Determination of Linear Viscoelastic Region (Strain Sweep)

1. Perform an oscillatory strain sweep at 1 Hz over 0.01–10% strain.
2. Identify the strain within which G′ and G″ and viscosity (η) remains constant (linear viscoelastic region, LVR).

Table 1
Physiochemical parameters of *Chlorella*-based drink during storage at 5 °C for 24 days

Parameters	Day 0	Day 7	Day 14	Day 21	Day 24
pH					
TA (g citric acid/100 g)					
SSC (°Brix)					
L*					
a*					
b*					
Apparent viscosity at 100 s^{-1} (Pa·s)					

3. Select a conservative working strain within the LVR (typical: 0.1%–1%) for any frequency measurements on subsequent tests.

3.7.4 Frequency Sweep

1. Conduct an oscillatory frequency sweep within the LVR using the selected strain.
2. Record G′ (storage) and G″(loss) across a frequency range of 100–0.2 Hz at 25 °C.

3.7.5 Viscosity Measurement (Flow Curve)

1. Switch to rotational mode and set a shear-rate ramp from 5 to 300 s^{-1}.
2. Record the corresponding shear stress (Pa)and calculate the apparent viscosity (Pa·s) at each shear rate.
3. Continue the measurement until the full shear-rate range is completed and save the viscosity curve data. Output: Viscosity vs Shear Rate curve (shear-thinning expected).

3.7.6 Completion

1. Lift the geometry, remove the sample, and clean the plates immediately.
2. Switch off the instrument if no further measurements are required.

3.7.7 Replicates and Data Handling

1. Perform three independent measurements for each sample. For each replicate, either:
   (a) Reload a fresh aliquot (recommended) or
   (b) Recondition the same sample with a standardized preshear and rest protocol (document precisely).
2. Document: For each replicate record sample ID, loading time, working strain, frequency sweep settings, shear protocol, and any warnings [4].

3.7.8 Key Formulas

1. Shear rate (plate–plate, approximate):

$$\dot{\gamma} = \frac{\Omega R}{h}$$

where Ω is angular velocity (rad·s^{-1}), $R$is plate radius (m), $h$is gap (m).

2. Shear stress (plate–plate):

$$\tau = \frac{2T}{\pi R^3}$$

where $T$is torque (N·m) and $\tau$is shear stress (Pa).

Table 2
Microbiological Quality (log CFU/g) During Storage [ISO 7218:2020]

Parameters	Day 0	Day 7	Day 14	Day 21	Day 24	Limit
TPC						<6.0
Psychrophilic						<5.0
Yeast and Mold						<3.0
E. coli						Absent
Salmonella						Absent
Listeria						Absent

3. Apparent viscosity:

$$\eta = \frac{\tau}{\dot{\gamma}} (\text{units Pa} \cdot \text{s})$$

4. Complex viscosity (from oscillatory data):

$$\eta^* = \frac{\sqrt{(G')^2 + (G'')^2}}{\omega}$$

where $\omega = 2\pi f(\text{rad}\cdot\text{s}^{-1})$, *f* is frequency (Hz).

3.8 Microbiological Analysis

1. Homogenize 10 g smoothie in 90 mL sterile peptone water (1: 10). Serial dilute 10-fold in Maximum Recovery Diluent.
2. Psychrophilic: Plate 1 mL on PCA, incubate 7 °C/10 days.
3. Yeast and Mold: Plate 1 mL on DRBC agar, incubate 25 °C/ 5 days (*see* **Note 5**).
4. Pathogens: *E. coli* (ISO 16649-2), *Salmonella* (ISO 6579-1), *Listeria* (ISO 11290-1). Absence in 25 g required (Table 2).
5. Count 30–300 CFU/plate colonies. Calculate $\log_{10}$ CFU/g [12].
6. Acceptance Criteria: TPC <10^6, Psychrophilic <10^5, Y + M <10^3 CFU/g; pathogens like *E. coli*: absent/10 g, others: absent/25 g.

3.9 Sensory Analysis

1. Sensory analyses follow international standards (ISO 8589: 2007).
2. Tests occur in a standard sensory room with ten individual taste booths with white lighting at 22 ± 2 °C
3. Serve 30-mL samples at room temperature (20 °C) in transparent plastic glasses coded with three random three-digit numbers.

Table 3
Sensory Evaluation Scores (n = 12 assessors, 9-point hedonic scale)

Attribute	Control Smoothie	*Chlorella* Smoothie
Appearance		
Flavor		
Texture		
Overall acceptability		
Defect scores (1–5 scale)		

4. Provide still mineral water and unsalted crackers as palate cleanser between samples (Table 3).
5. Recruit 12 trained assessors (6 women, 6 men; 22–68 years) screened for sensory acuity (color, odor, taste, texture).
6. Present samples monadically (control + *Chlorella* smoothie, balanced order in 3 replications).
7. Score on 9-point hedonic scale based on visual appearance, flavor, texture, overall acceptability (1 = dislike extremely; 9 = like extremely).
8. Score defect intensity on 5-point scale: off-color, off-odors, lumpiness, turbidity, phase separation (1 = none; 5 = extreme) (*see* **Note 6**).
9. Analyze data by mean ± SD, ANOVA, Tukey's HSD ($p < 0.05$); minimum acceptability ≥6.0/9 overall liking [13, 14].

3.10 Statistical Analysis

1. The experiments should be conducted in triplicates and the results are represented as mean ± SD.
2. Use SPSS, R, or GraphPad Prism 7.0 software for the statistical analysis.
3. One-way ANOVA is carried out to determine significant differences among sample means. Post hoc comparisons are performed using Tukey's Honest Significant Difference (HSD) test. Statistical significance is considered at a p-value less than 0.05

4 Notes

1. **Smoothie Preparation**
   (a) Phase separation: Increase Ultra-Turrax homogenization to 10,000 rpm/3 min post-blending; add 0.1–0.2% xanthan gum if persistent

Fig. 1 Preparation steps of green *Chlorella* smoothie

(b) *Chlorella* fishy odor: Use broken cell wall powder; mask with pineapple juice (10%) or ginger (2 g/100 g)

(c) Green lumps: Pre-hydrate *Chlorella* powder in 20 mL warm water (40 °C) 15 min before blending.

(d) This smoothie formulation can be modified by incorporating alternative fruit- or plant-based ingredients, as demonstrated in previous studies that developed *Chlorella*-based beverages and value-added products (Fig. 1) such as the following:

Formulation	Ingredients and quantities (per 100 mL)	*Chlorella* powder	Reference
Green apple ginger smoothie	Green Apple 40 g Ginger 5 g Water 50 mL	10 g	[15]
Coca Functional drink	Coca base 95 mL	*Chlorella* hydrolysate 1–3% (w/v)	[6]
Fermented soy drink	Soy beverage 98.5 mL Xylitol 1.5 g	1.5% (w/v)	[4]
Apple–Pear Microalgae Smoothie	Apple pulp 40 g Pear pulp 40 g Water/fruit juice 20 mL	3% w/v	[14]

2. **Physicochemical Analysis**
   (a) Refractometer SSC reading high: Wipe prism thoroughly; calibrate with distilled water between samples.
   (b) TA titration endpoint unclear: Use fresh 0.1% phenolphthalein; gentle magnetic stirring prevents CO_2 absorption.
3. **Rheology**
   (a) Cover the edge with light paraffin oil, if sample dries.
   (b) Preequilibrate for 15 min at 20 °C before measurement.
   (c) Keep shear rate <500 s^{-1}; measure immediately after loading because viscosity drops 20% in 5 min
4. **Color Measurement**
   (a) a* >2 units: Chlorophyll degradation, hence analyze within 2 h of postpreparation.
   (b) $\Delta E > 5$ (browning): verify blanching 95 °C exactly 2 min.
5. **Microbiological Analysis**
   (a) Use laminar flow hood during packaging.
   (b) **Psychrophilic overgrowth**: Incubate PCA exactly 7 °C (not 4 °C); count after 10 days minimum.
   (c) **Yeast/mold colonies spreading**: Add chloramphenicol (0.05 g/L) to DRBC agar.
6. **Sensory Evaluation:**
   (a) Low flavor score (<6.0/9): Screen assessors for *Chlorella* tolerance first, serve chilled (8 °C) not room temperature
   (b) Panel fatigue: maximum 3 samples/session, 10 min palate cleansing with water and cracker.

References

1. Corbo MR, Bevilacqua A, Petruzzi L, Casanova FP, Sinigaglia M (2014) Functional beverages: the emerging side of functional foods. Compr Rev Food Sci Food Saf 13:1192–1206. https://doi.org/10.1111/1541-4337.12109
2. World Health Organization. Regional Office for Europe (2021) Plant-based diets and their impact on health, sustainability and the environment: a review of the evidence: WHO European Office for the Prevention and Control of Noncommunicable Diseases. World Health Organization. Regional Office for Europe, Copenhagen
3. Caporgno MP, Mathys A (2018) Trends in microalgae incorporation into innovative food products with potential health benefits. Front Nutr 5:58. https://doi.org/10.3389/fnut.2018.00058
4. Csatlos NI, Simon E, Teleky BE, Szabo K, Diaconeasa ZM, Vodnar DC, Ciont Nagy C, Pop OL (2023) Development of a fermented beverage with *Chlorella vulgaris* powder on soybean-based fermented beverage. Biomolecules *13*(2):245. https://doi.org/10.3390/biom13020245
5. de Andrade LM, de Andrade CJ, Dias M, Ao Nascimento C, Mendes MA (2018) Chlorella and Spirulina microalgae as sources of functional foods, nutraceuticals, and food supplements; an overview. MOJ Food Process Technol *6*(1):1–14. https://doi.org/10.15406/MOJFPT.2018.06.00144

6. Dian I, Harifa PS, Bustari H, Trisla W, Zahtamal AK, Hannan SUR (2025) Diversification of cocoa drinks with Chlorella vulgaris hydrolysate as a functional innovation for health benefits. Asian J Dairy Food Res 44(4):594–600. https://doi.org/10.18805/ajdfr.DRF-488
7. Castillejo N, Martínez-Hernández GB, Hernández-Carrión M, Gómez PA, Artés F, Artés-Hernández F (2018) Natural vitamin B12 and fucose supplementation of green smoothies with edible algae and related quality changes during their shelf life. J Sci Food Agric *98*(6):2411–2421. https://doi.org/10.1002/jsfa.8733
8. Castillejo N, Martínez-Hernández GB, Gómez PA, Artés F, Artés-Hernández F (2016) Red fresh vegetables smoothies with extended shelf life as an innovative source of health-promoting compounds. J Food Sci Technol *53*(6):2748–2759. https://doi.org/10.1007/s13197-015-2143-2
9. Association of Official Analytical Chemists (AOAC) (2005) Official methods of analysis, 18th edn. AOAC International. Method 981.21: pH of acidified foods
10. Association of Official Analytical Chemists (AOAC) (2005) Official methods of analysis, 18th edn. AOAC International. Method 942.15: Acidity (titratable) of fruit products
11. Bustamante ML, Góngora JD, Drider D (2023) Development of a fermented beverage with *Chlorella vulgaris* and probiotic strains: effects on physicochemical, microbiological, and antioxidant properties. Appl Sci 13(3): 1473. https://doi.org/10.3390/app13031473
12. International Organization for Standardization (2020) ISO 7218:2020 microbiology of the food chain – general requirements and guidance for microbiological examinations. ISO. https://www.iso.org/standard/72118.html
13. International Organization for Standardization (2007) ISO 8589:2007 sensory analysis – general guidance for the design of test rooms. https://www.iso.org/standard/36385.html
14. Viegas Â, Raio M, Alegria MJ, Raymundo A (2025) Upcycling apple and pear pulp into functional smoothies: a sustainable approach using microalgae and antioxidant enhancement. Food Biosci 71:107241. https://doi.org/10.1016/j.fbio.2025.107241
15. Kumar D (2017) Value addition of green apple ginger smoothie by enrichment with *Chlorella vulgaris*: sensory evaluation. Int J Appl Home Sci 4(9 & 10):800–804

Chapter 34

Preparation, Nutritional and Sensory Analysis of *Chlorella*-Based Cookies

Shrika Kumaresan and Dhanasekaran Dharumadurai

Abstract

Chlorella is a green, unicellular eukaryotic microalgae, rich in proteins and micronutrients, that has gained popularity as a functional ingredient in food products including bakery products like cookies. This chapter describes a detailed protocol for preparing *Chlorella*-based cookies using wheat and millet flour blends to develop nutrient-dense, ready-to-use therapeutic food (RUTF)-like formulations for the health-conscious and diabetic populations. The formulation uses low levels of *Chlorella* powder to enhance protein, mineral, and bioactive content while maintaining desirable cookie texture and crispness. The physicochemical and quality analysis include proximate composition, water activity (a_w), color measurement, in vitro digestibility and sensory evaluation. This protocol can be modified using different cereal bases and inclusion levels of *Chlorella*, making it easier to produce *Chlorella*-enriched cookies as a healthy snack and RUTF options on an industrial scale.

Key words *Chlorella vulgaris*, Microalgae fortification, Algae-based cookies, Functional food, Sensory evaluation

1 Introduction

Microalgae are gaining more attention as sustainable sources of nutrients and bioactive compounds that can be used in the production of functional foods and nutraceuticals. High concentrations of protein, essential fatty acids, vitamins, minerals, polysaccharides, pigments, and phenolic compounds are provided by a number of species, including *Spirulina*, *Chlorella*, and other green microalgae. This supports their use as fortifiers in a variety of food matrices, including snacks, drinks, and baked goods [1]. Among these, *Chlorella vulgaris* is one of the most widely studied edible microalgae, containing 30 to 70% protein on a dry basis. It has substantial amount of carbohydrates, lipids, chlorophyll, carotenoids, vitamins, and its essential amino acid profile is similar to high-quality conventional protein sources This composition supports several health effects, including antioxidant, immunomodulatory,

Dhanasekaran Dharumadurai (ed.), *Chlorella Genomics, Cultivation, and Bioenterprise: Green Economy through Algal Innovations*, Springer Protocols Handbooks, https://doi.org/10.1007/978-1-0716-5276-3_34,

© The Author(s), under exclusive license to Springer Science+Business Media, LLC, part of Springer Nature 2026

cholesterol-lowering and glycemic-regulating activities, as well as potential benefits in sports nutrition. *C. vulgaris* has been granted "generally recognized as safe" (GRAS) status for usage in specific foods, which has facilitated its incorporation into products such as pasta, cookies, breads and dairy or plant-based beverages [2].

Despite these benefits, there are several challenges in the way of *Chlorella*'s widespread use in cookies and other baked goods. At higher *Chlorella*-inclusion levels, consumers may be less receptive due to the green hue, fishy flavor, and odor. Hence, it requires careful formulation with correct dose adjustment, flavor masking, and matrix selection. Additionally, the presence of microalgal biomass can affect dough rheology, spread ratio, texture, hardness, and water content [3]. These characteristics must be optimized to preserve the intended taste and shelf life of cookies. Pigment stability, antioxidant retention, and structure can be impacted by variables including mixing, baking temperature and duration, and *Chlorella* powder particle size, which can affect both appearance and nutritional delivery. Hence, addressing these technical and sensory challenges is crucial to ensure that nutritional benefits do not compromise product quality.

This chapter describes a practical guide for making nutritious *Chlorella* cookies using millet and whole wheat flour, covering key ingredients selection, dough preparation, baking conditions, and post-baking handling. It also describes methods for measuring the proximate composition, water activity, color, texture profile analysis, and sensory qualities [4, 5]. This information supports the creation of *Chlorella*-enriched cookies that serve as functional snacks and similar RUTF-like products made from local grains. This framework is easy to modify for new ideas, like adding other varieties of flour and beneficial ingredients like probiotics, to support the evolving landscape of microalgae-based functional bakery products.

2 Materials

2.1 Cultivation and Biomass Production of *Chlorella vulgaris*

1. Microalgal culture.
2. BG 11 media.
3. Erlenmeyer flasks.
4. Fine cotton mesh cloth.
5. Distilled water.
6. Weigh balance.
7. Hot air oven.
8. Lyophilizer.

2.2 Cookie Preparation

1. Whole wheat flour or millet flour.
2. Sugar (white/brown).
3. Fat (butter/oil).
4. Baking soda or baking powder.
5. Eggs or curd.
6. Milk.
7. Vanilla extract.
8. Salt.
9. Baking trays.
10. Microwave Oven.

2.3 Proximate Composition

2.3.1 Moisture Analysis

1. Moisture Analyzers (METTLER TOLEDO HC103 Halogen Moisture Analyzer).
2. Aluminum sample pans (Ø 70 mm).
3. Hot air oven.

2.3.2 Ash Content Determination

1. Muffle furnace.
2. Porcelain crucibles.

2.3.3 Protein (Kjeldahl)

1. Burette (50 mL, 0.1 mL).
2. Kjeldahl flasks (500 mL).
3. Conc. H_2SO_4 (98%).
4. Kjeldahl catalyst tablets (K_2SO_4 and $CuSO_4$).
5. NaOH (40% w/v).
6. Boric acid (4%w/v).
7. Tashiro's indicator (or 0.1% methyl red and methylene blue).
8. 0.1 N HCl.

2.3.4 Fat (Soxhlet)

1. Petroleum ether (40–60 °C).
2. Thimble paper.

2.3.5 Colorimetric Analysis (CIELAB)

1. Colorimeter (Konica Minolta CR-400 or equivalent).
2. White calibration tile.
3. Glass reference plate.

2.3.6 Texture Profile Analysis (TPA)

1. Texture analyzer (TA.XT Plus or Stable Micro Systems).
2. TPA probe (P/36R, 5 kg load cell).
3. Aluminum compression plates.

2.3.7 In Vitro Digestibility

1. DNS reagent (3,5-Dinitrosalicylic acid).
2. Glucose standard.
3. White bread (reference).
4. Pancreatin (2.5% in 0.1 M malate buffer pH 6.0).
5. Pepsin (1 g/10 mL 0.05 M HCl).
6. Amyloglucosidase (0.1 mL).

2.3.8 Sensory Analysis

1. Mineral water.
2. Toothpicks, expectoration cups.
3. 9-point hedonic scale ballots.

2.3.9 General Lab Consumables

1. Analytical balance (±0.1 mg).
2. Mortar/pestle (80 mesh sieving).
3. Centrifuge.
4. Vortex mixer.
5. Magnetic stirrer.
6. Water bath (37 °C, 95 °C).
7. Pipettes (10–1000 μL).
8. Desiccators (silica gel).
9. Gloves.

3 Methods

3.1 Cultivation and Biomass Production of Chlorella vulgaris

1. Prepare BG11 medium using distilled water or RO water, and adjust pH to about 6.5–7.5.
2. Inoculate 5% (v/v) of *Chlorella vulgaris* into sterilized medium under aseptic conditions.
3. Incubate the culture for 7–14 days in transparent containers such as Erlenmeyer flasks, plastic bottles or photobioreactors. Maintain continuous aeration and illumination at 1000–3000 lux (cool white fluorescent light or LED). Keep the temperature at 22 °C ± 2 °C.
4. Once culture attains dense green coloration indicating high biomass, harvest by filtration using fine mesh (10–20 μm mesh) or filter cloth [6].
5. Rinse the harvested biomass with distilled water to remove any culture medium residues if required and spread the wet biomass in thin layers on trays for drying.
6. Use freeze-drying (lyophilization) for best preservation of bioactive compounds. Alternatively, drying can be performed in a hot air oven at controlled temperature of 40 to 50 °C to limit the degradation of nutrients and pigments.

7. Dry until the moisture content is below 10% and store dried biomass in airtight containers away from light, heat, and moisture [4].

3.2 Cookie Preparation

1. Substitute 6 % of the total flour weight with *Chlorella* powdered meal (the level shown to give no significant sensory difference and acceptable texture) (*see* **Note 2**).
2. Standard Measurements for Cookie Dough using 100 g Flour:
   (a) Flour (wheat or millet): 100 g.
   (b) *Chlorella* powder: 2–6 g (2–6% of flour weight; adjust based on desired fortification).
   (c) Sugar: 50–60 g.
   (d) Fat (butter/oil): 50–70 g.
   (e) Egg or thick curd: 30–40 g (approximately 1 small egg).
   (f) Milk or plant-based milk: 30–50 mL.
   (g) Baking powder: 1–2 g.
   (h) Salt and vanilla extract: small teaspoon (as per taste).
   (i) Maltodextrin (Optional): 2 g (Improves dough handling at 6 % algae).
   (j) These measurements typically make about 12–15 medium-sized cookies (around 10–12 g dough per cookie).
3. Mix dry ingredients (flour, *Chlorella* meal, baking powder, powdered sugar, salt) thoroughly.
4. Add wet ingredients like melted butter, egg, or curd, and vanilla extract, then incorporate the dry blend.
5. Preheat oven to 170 °C (*see* **Note 1**).
6. Knead manually until a homogeneous dough forms. Sheet the cookie dough with a rolling pin and its shape is cut using circular disks (5 mm thickness).
7. The controls used for the analysis is made from wheat and millet flour devoid of microalgae.
8. Line baking trays with parchment paper and bake for 12–15 min, or until cookies are lightly golden on edges (Fig. 1).
9. The cookie samples were cooled to room temperature (25 ± 2 °C) and stored in airtight containers for further analysis (*see* **Note 6**) [5].

3.3 Physiochemical Analysis

3.3.1 Proximate Composition

1. **Moisture Content**
   (a) Grind cookies to uniform powder (*see* **Note 8**).
   (b) Weigh 2–3 g sample into aluminum pan (*see* **Note 3**).

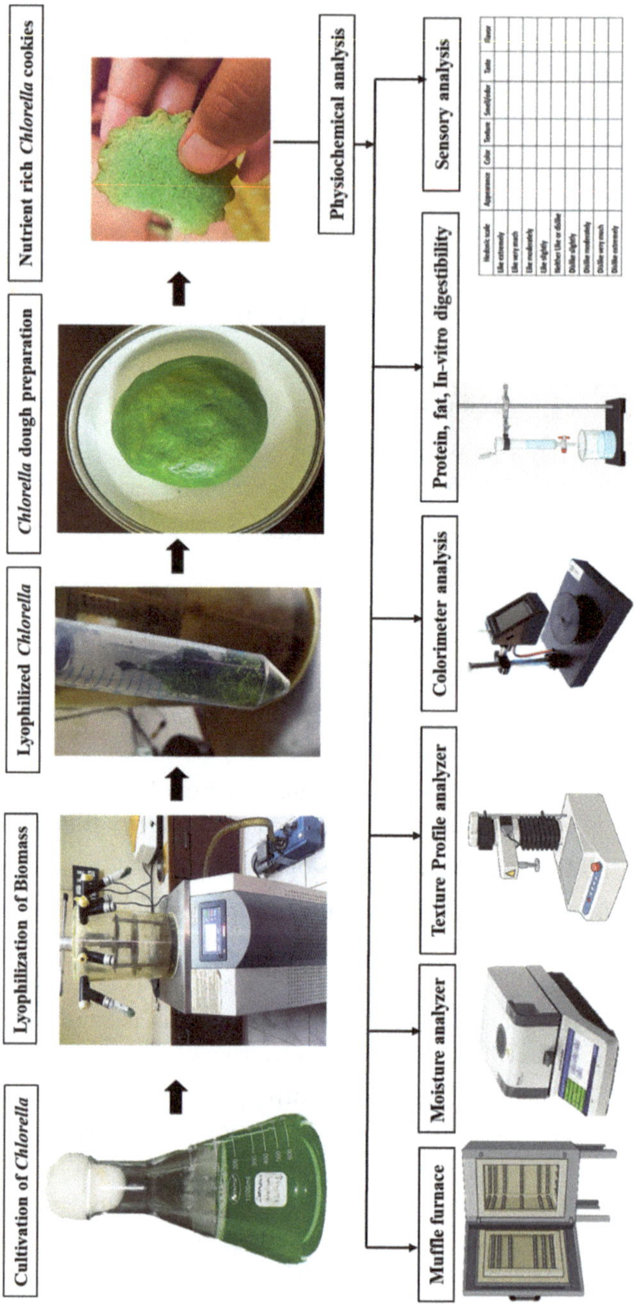

Fig. 1 Preparation and quality analysis of *Chlorella*-based cookies

(c) Dry at 60–75 °C for 10–15 min in moisture analyzer or 75 °C for 3 h in hot air oven until constant weight [7].

(d) Calculate moisture (%) = (loss in weight/initial weight) × 100.

2. **Ash Content**

(a) Preignite empty crucibles at 550 °C, cool in desiccator, and record weight (W_c).

(b) Add 2–5 g ground cookie sample and record crucible + sample weight ($W_c + s$).

(c) Incinerate in muffle furnace at 550 °C for 4–6 h until residue becomes light grey/white.

(d) Cool in desiccator and record crucible + ash weight ($W_c + a$).

(e) Repeat heating if residue appears dark [8].

- Sample weight $= W_s = W_c + s - W_c$
- Ash weight $= W_a = W_c + a - W_c$

$$\%\text{Ash} = \frac{W_a}{W_s} \times 100$$

3. **Protein**

(a) Grind cookies to fine powder (<0.5 mm) and weigh 1 g of sample into a Kjeldahl digestion flask.

(b) Add 20 mL conc. H_2SO_4 and 1 Kjeldahl tablet (10 g K_2SO_4 + 0.5 g $CuSO_4$)

(c) Heat at 370–410 °C for 2–3 h until clear green solution (no char)

(d) Cool the digest, then dilute to 100 mL with distilled water

(e) Transfer digest to distillation unit

(f) Add 40–50 mL of 40% sodium hydroxide (NaOH) solution carefully to make the digest alkaline (releases NH_3)

(g) Distill 150 mL (5–7 min) into flask containing 25 mL 4% boric acid with 3 drops Tshiro's indicator

(h) Titrate with 0.1 N HCl to gray–pink endpoint (VS mL), Run blank (VB mL).

(i) Calculate nitrogen content from titration volume; multiply by 6.25 to get protein content [9].
Nitrogen (%)

$$\%\text{N} = \frac{\left(V_{\text{sample}} - V_{\text{blank}}\right) \times N_{\text{acid}} \times 14.007}{\text{Sample mass (g)}}$$

$$\%\text{Protein} = \%\text{N} \times \text{conversion factor (6.25)}$$

4. **Fat (Soxhlet Extraction)**
   (a) Weigh 2–5 g of dried and ground biomass into a cellulose extraction thimble.
   (b) Place the thimble in the Soxhlet extractor.
   (c) Add approximately 150–200 mL of petroleum ether with a boiling range 30–60 °C, as solvent into a round bottom flask.
   (d) Extract continuously for 4–6 h (or 16 h for thorough extraction) with a condenser to reflux solvent.
   (e) After extraction, remove solvent by evaporation in a drying oven at 100–105 °C for 30 min to evaporate residual solvent.
   (f) Cool the flask in a desiccator and weigh to determine extracted fat.
   (g) Calculate fat content as percentage of initial dry sample weight.
   Fat content (%) formula:

$$\text{Fat\%} = \frac{(W_2 - W_1) \times 100}{W_0}$$

where

W_0 = Weight of dry sample (g) (typically 2–5 g)
W_1 = Weight of empty flask/beaker (g)
W_2 = Weight of flask + extracted fat residue (g)

5. **Carbohydrates**

   Calculated by difference in proximate analysis [10].

   Carbohydrates (%) = 100 − (Moisture + Ash + Protein + Fat).

3.3.2 Texture Profile Analysis (TPA)

1. Use a texture analyzer with a flat cylindrical probe (e.g., 25-mm diameter).
2. Place the freshly baked cookie (usually within 24 h) sample on the lower aluminum plate platform (*see* **Note 4**).
3. Set test parameters:
   (a) Pretest speed: 1 mm/s
   (b) Test speed: 1 mm/s (compression)
   (c) Post-test speed: 1 mm/s
   (d) Compression distance: 25–50% of sample thickness (usually 2–3 mm for cookies)
   (e) Time between compressions: 5 s

4. Perform two consecutive compressions imitating two bites ("two-bite test").
5. Record force vs. time curves [11].
6. Analyze values from triplicate cookies.
7. Calculation of texture parameters:

$$\text{Hardness} = \text{Peak force during compression (g or N)}$$

$$\text{Cohesiveness} = \frac{\text{Area under 2nd compression curve}}{\text{Area under 1st compression curve}}$$

$$\text{Springiness} = \frac{\text{Distance recovered during 2nd compression}}{\text{Distance compressed during 1st compression}}$$

$$\text{Chewiness} = \text{Hardness} \times \text{Cohesiveness} \times \text{Springiness}$$

3.3.3 Color Measurement (CIE Lab*)

1. Place a flat, crack-free cookie surface (on both top and bottom) on a white or black background. Avoid gloss/shadows and do 5 replicates (*see* **Note 4, 6**).
2. Calibrate with white standard plate ($L^* = 100$) and black ($L^* = 0$) standards.
3. Place the instrument's measurement aperture flush against the cookie surface to prevent external light interference. Use standard conditions (D65 illuminant, 2° observer) (*see* **Note 5**).
4. Position the probe (D65 illuminant, 10° observer) perpendicular to the surface.
5. Record the L* (lightness), a* (red–green), and b* (yellow–blue) values for each sample and for the control cookies.
6. Measure each cookie at three to five different random points.
7. Compute the overall color difference between the control and *Chlorella*-enriched cookies using the formula:

$$\Delta E^* = \sqrt{\left[(L_1 - L_2)^2 + (a_1 - a_2)^2 + (b_1 - b_2)^2\right]},$$

where subscript 1 represents the control and 2 represents the treated sample. $\Delta E > 2$ indicates a noticeable color difference. The characteristic greenish hue reflects more negative a* values [12].

3.3.4 In Vitro Digestibility

1. Weigh 0.5 g ground cookie powder into 50 mL tube. Add 30 mL RO water. Stir at 37 °C for 10 min (Table 1).
2. Add 1 mL pepsin solution. Incubate 37 °C, 30 min. $T = 0$: Withdraw 1 mL aliquot, add 4 mL 80% ethanol to stop reaction.
3. Add 0.1 mL amyloglucosidase + 5 mL pancreatin solution. Incubate 37 °C with stirring.

Table 1
Physicochemical and digestibility parameters of *Chlorella*-based cookies

Parameter	Control	6% *Chlorella*
Proximate (9 g/100 g)		
Moisture		
Protein		
Fat		
Ash		
CIELAB color		
L*		
a*		
b*		
TPA texture		
Hardness (*N*)		
Chewiness (*N*)		
Digestibility		
HI (%)		
eGI		

4. Withdraw 1 mL aliquots at 20, 60, 120 min. Stop each with 4 mL 80% ethanol.
5. Glucose Quantification: Centrifuge aliquots (3000 rpm, 10 min). Use DNS (3,5-Dinitrosalicylic acid) method on supernatant.
6. Glucose (mg/g) = [Absorption sample/Absorption standard] × Standard concentration × Dilution factor
7. Hydrolysis Index (HI): Plot glucose release vs. time. Calculate AUC (area under curve) for sample vs. reference (white bread = 100)

$$\text{HI}\ (\%) = (\text{AUC sample}/\text{AUC reference}) \times 100$$

8. Calculate the estimated Glycemic index (eGI) from the Hydrolysis Index (HI) obtained from in vitro starch digestibility test [5].

$$eGI = 39.71 + 0.549 \times HI$$

3.3.5 Sensory Evaluation

1. Recruit 60–100 consumers from target population (Age 18–65 years, cookie eaters)(*see* **Note 9**).
2. Prepare samples and code with random 3-digit numbers.
3. Set up separate tasting booths with neutral lighting and no interactions between consumers.
4. Serve three to six samples in a random order each session.
5. Instruct consumers to rate overall liking on a 9-point scale (1 = dislike extremely, 9 = like extremely).
6. After tasting and rating each sample, the customers should rinse their mouth with water in between.
7. Collect all the data and evaluate acceptance and preference using pairwise tests and ANOVA [13].

3.3.6 Statistical Analysis

1. All the experiments should be conducted in triplicates and represented as mean ± SD in results.
2. For statistical analysis, use GraphPad Prism 7.0, R, or SPSS.
3. To find significant differences between sample means, one-way ANOVA is used. Post hoc comparisons are performed using Tukey's Honest Significant Difference (HSD) test.
4. Statistical significance is considered at a p-value less than 0.05.

4 Notes

1. Always maintain baking temperature at 110–170 °C, higher temperatures (>180 °C) can intensify brown color and causes pigment degradation. It may produce bitter taste and make cookies look dark green–brown in color.
2. Do not prepare dough with more than 6% of *Chlorella* without prior testing, because higher levels (>8%) affect taste and color.
3. Include a control cookie (without *Chlorella*) for all physio-chemical analysis like color ΔE, texture, digestibility, sensory tests.
4. When measuring color or texture, avoid using metal trays. They reflect light and alter readings. Use nonreflective, matte surfaces.
5. In colorimetric analysis, measure center 3 cm^2 flat areas only, cracks/edges give false L^* values.
6. Do not expose cookies to ambient light before color measurement, surface oxidation changes L*, a*, b* values. Therefore, measure immediately after cooling.
7. Store the cookies in desiccator after baking, humidity alters a* greenness within 2 h.

8. High variability in moisture results is caused by inconsistent grinding, use an 80 mesh sieve for uniformity.
9. Sensory analysis requires institutional ethical clearance ($n = 80$ consumers).

References

1. Heo Y, Kim M-Y, Cho JY (2025) Chlorella vulgaris, a representative edible algae as integrative and alternative medicine. Integr Med Res. https://doi.org/10.1016/j.imr.2025.101147
2. Mendes AR, Lordelo M, Prates JAM, Spínola MP (2024) Chemical compounds, bioactivities, and applications of Chlorella vulgaris in food. Feed and medicine. Appl Sci 14(23):10810. https://doi.org/10.3390/app142310810
3. Uribe-Wandurraga ZN, Igual M, Martínez-Monzó J, García-Segovia P (2019) Effect of microalgae addition on mineral content, colour and mechanical properties of breadsticks. Food Funct 10(8):4685–4692. https://doi.org/10.1039/c9fo00286c
4. Dündar AN, Şahin OI, Sarıcaoğlu FT (2023) Low-fat cookies with *Chlorella vulgaris*: effects on dough rheology, physical, textural and sensory properties of cookies. Gida 48(3):526–544. https://doi.org/10.15237/gida.GD22076
5. Selvaraju K, Mathur M, Satheeshkanth SSM, Sahu JK, Hariprasad P, Malik A (2024) Deciphering the effect of *Chlorella minutissima* incorporation in gluten-free millet cookies: physical and functional properties. Algal Res. https://doi.org/10.1016/j.algal.2024.103295
6. Jui TJ, Tasnim A, Islam SMR, Manjur OHB, Hossain MS, Tasnim N, Karmakar D, Hasan MR, Karim MR (2024) Optimal growth conditions to enhance *Chlorella vulgaris* biomass production in indoor phyto tank and quality assessment of feed and culture stock. Heliyon 10(12):Article e32731. https://doi.org/10.1016/j.heliyon.2024.e32731
7. AOAC International (2019) Official methods of analysis of AOAC International. Method 925.10: Solids (total) and loss on drying (moisture) in flour, 21st edn. AOAC International
8. AOAC International (2019) Official methods of analysis of AOAC international. Method 923.03: ash of flour, 21st edn. AOAC International
9. AOAC International (2019) Official methods of analysis of AOAC international. Method 2001.11: protein (crude) in animal feed, forage (plant tissue), grain, and oilseeds, 21st edn. AOAC International
10. FAO/INFOODS (2012) FAO/INFOODS guidelines for checking food composition data prior to the publication of a user table/database—Version 1.0. Food and Agriculture Organization of the United Nations. https://www.fao.org/fileadmin/templates/food_composition/documents/Guidelines_data_checking_final_oct2012.pdf
11. AACC International (2010) Approved methods of analysis, 11th edn. Method 74-09.01
12. International Organization for Standardization (2008) ISO/CIE 11664-4:2008 – Colorimetry – Part 4: CIE 1976 Lab colour space*. ISO. https://www.iso.org/standard/52497.html
13. International Organization for Standardization (2014) ISO 11136:2014. Sensory analysis – methodology – general guidance for conducting hedonic tests with consumers in a controlled area. ISO Standard No. 11136:2014. https://www.iso.org/standard/50125.html

Chapter 35

Microencapsulation of *Chlorella* for Sensory Improvement

Shrika Kumaresan and Dhanasekaran Dharumadurai

Abstract

A microalga such as *Chlorella* vulgaris has intense color, fish-like smell, and taste, which frequently restrict its use in food matrices. Despite its substantial nutritional and health benefits, these factors may reduce consumer acceptance. The best way to reduce these sensory problems and protect sensitive pigments and bioactive compounds during processing and storage is through microencapsulation. Encapsulation systems are designed to improve handling, control the release of pigments and antioxidants in the digestive tract, and increase their stability. This chapter outlines a detailed procedure for making multicomponent spray-dried microcapsules that contain *Chlorella* extracts.

Key words Microencapsulation, Sensory improvement, *Chlorella vulgaris*, Spray-drying, Biopolymer wall materials

1 Introduction

Chlorella vulgaris, a unicellular green microalga, commonly found in aquatic system is recognized for its rich content of proteins, lipids, vitamins, minerals, and compounds such as lutein and antioxidants. Despite being acknowledged as a nutrient-dense microalga, its strong green color, herbal or marine-like flavor, and distinctive odor often prevent it from being directly incorporated into food products. These sensory characteristics are frequently seen as off-notes in mainstream foods, particularly at inclusion levels necessary to achieve meaningful nutritional or functional claims, but they can be useful in some niche products. *Chlorella* pigments and bioactive compounds are also susceptible to light, oxygen, and processing conditions, which can cause degradation that can further change color and flavor while reducing health-related functionality [1]. Hence, there is a need for technological solutions that allow the benefits of *Chlorella* to be delivered without sacrificing sensory quality.

Among many microencapsulation methods like complex coacervation, extrusion-based techniques, spray-drying is a widely used

Dhanasekaran Dharumadurai (ed.), *Chlorella Genomics, Cultivation, and Bioenterprise: Green Economy through Algal Innovations*, Springer Protocols Handbooks, https://doi.org/10.1007/978-1-0716-5276-3_35,

© The Author(s), under exclusive license to Springer Science+Business Media, LLC, part of Springer Nature 2026

method due to its scalability, cost-effectiveness, and ability to produce shelf-stable dry powders with uniform particle size [2]. The type of wall material chosen also plays an important role in determining the dry powder morphology, the release of core material compounds, and overall encapsulation efficiency (EE). Combinations of biopolymers like lecithin, prebiotic fibers such as inulin and modified starch like octenyl succinic anhydride starch can successfully encapsulate the *Chlorella* extract. From the sensory perspective, *Chlorella* encapsulation has many advantages. Firstly, the off-flavors and marine-like odor of *Chlorella* is diminished by delayed release of the volatile compounds into the surrounding food environment. Secondly, by dispersing the pigment molecules inside the microcapsules, it reduces the visual impact of pigments that reduces the intensity of overall green coloration while still delivering bioactive compounds. Thirdly, this method allows uniform distribution of tiny particles rather than coarse aggregates or lumps, thereby enhancing the texture and mouthfeel. This uniquely designed microcapsules are important for fortifying beverages, cookies, or soft bakery food preparation [3].

This chapter describes a detailed procedure for preparing microencapsulated *Chlorella vulgaris* using spray-drying technique with these multicomponent carriers. It contains extraction of bioactives, feed emulsion formulation, spray-drying parameters optimization, and post-process characterization such as Encapsulation efficiency rate, particle size distribution, flowability, dissolution rate, and sensory analysis [3]. This approach facilitates the development of functional foods like bakery products, drinks that maintain the health advantages of *Chlorella* without losing stability and sensory appeal.

2 Materials

2.1 Chlorella *Extract Preparation*

1. *Chlorella vulgaris* lyophilized powder.
2. Ultra sonicator (VCX 500, Sonics & Materials Inc., USA).
3. Ethanol (96%, analytical grade).
4. Vacuum rotary evaporator (Buchi R-100, Switzerland).
5. UV–Vis spectrophotometer.

2.2 Preparation of Multicomponent Wall Material

1. Maltodextrin DE 10–12.
2. Modified starch (OSA-MS).
3. Inulin.
4. Soy lecithin.
5. Ultra-Turrax T25 Homogenizer.
6. Dynamic light scattering (DLS, Zetasizer Nano ZS, Malvern).

2.3 Spray-Drying

1. Buchi B-290 Mini Spray dryer.
2. Peristaltic pump (feed rate 5–15 mL/min).
3. Nozzle (0.7 mm diameter, two-fluid).
4. Cyclone separator.
5. Inlet air heater (150–190 °C).

2.4 Physiochemical Characterization of Microcapsules

2.4.1 Encapsulation Efficiency

1. Cold hexane.
2. 0.01% Butylated Hydroxytoluene (BHT).
3. Acetone.

2.4.2 Particle Morphology

1. Spray-dried powder samples.
2. Double-sided carbon tape.
3. Aluminum SEM stubs.
4. Gold sputter coater (Quorum Q150R ES or equivalent).
5. Scanning Electron Microscope (JEOL JSM-6390LV).
6. Tweezers (anti-magnetic, fine tip).
7. Compresses nitrogen gas.
8. Desiccator (with P_2O_5 desiccant).

2.4.3 Particle Size

1. Laser diffraction particle sizer (Malvern Mastersizer 3000).
2. Compressed air (2–4 bar).
3. Dispersion unit (Scirocco 2000, 50% feeder speed).

2.4.4 Flowability Analysis

1. Tapped density tester (Erweka SVM 122, Germany).
2. Sieve shaker (Retsch AS 200, 500 μm mesh).

2.4.5 Dissolution Testing

1. Syringe filters (0.45 μm PTFE, 25 mm).
2. 1-cm quartz cuvettes.
3. Folin–Ciocalteu reagent (2N, diluted 1:10).
4. Na_2CO_3 (7.5 % w/v).
5. Gallic acid standard (0–500 μg/mL).

2.4.6 Sensory Analysis

1. Trained panelists ($n = 12$, 6M/6F, 22–50 years).
2. Transparent plastic cups (30 mL, odorless).
3. Mineral water (still), unsalted crackers.
4. Sensory scorecards (9-point hedonic + 5-point defect scales).
5. Individual sensory booths (white lighting, 22 ±2 °C, 50–60% RH).

2.4.7 General Laboratory Consumables

1. Magnetic stirrer hotplate (IKA C-MAG HS 7, 100 rpm).
2. Dissolution vessels (50 mL, 100 mL beakers, borosilicate).
3. Thermometer (25 ± 0.1 °C).
4. Vortex.
5. Water bath.
6. Analytical balance (0.001 g precision).
7. Graduated cylinder (50 mL, 100 mL borosilicate glass).
8. Falcon tubes.
9. Micropipettes (200 μL, 1 mL).
10. Tips.
11. Pasteur pipettes (10 mL).
12. Cooling centrifuge.
13. Distilled water (5 L).

3 Methods

3.1 **Chlorella** *Extract Preparation*

1. Suspend 50 g of *C. vulgaris* biomass powder in 1 L of 70% ethanol (1:20 w/v).
2. Sonicate the suspension using Ultra sonicator (VCX 500, Sonics & Materials Inc., USA) operating at 500 W power, 20 kHz frequency, and 40 kHz probe amplitude for 30 min at 25 ± 1 °C, to prevent ethanol evaporation.
3. Centrifuge the extract at 10,000 rpm for 15 min at 4 °C.
4. Concentrate the extract to 20% solids (w/v) using rotary vacuum evaporator (Büchi R-100, Switzerland) at 40 °C bath temperature and 100 mbar vacuum pressure.
5. Verify the final extract volume (150 mL) containing 5–8% total carotenoids and 15–20% proteins, using UV–Vis spectrophotometry at 470 nm and Bradford assay, respectively [3].

3.2 Preparation of Multicomponent Wall Material

1. These wall materials are sequentially dissolved in 1.5 L distilled water and heated to 40 °C under constant stirring in a magnetic stirrer (500 rpm, IKA RCT basic) to prepare a carrier solution with 20% total solids (w/w).
2. Firstly, add maltodextrin (250 g) and allow to completely dissolve for 30 min.
3. Followed by OSA-MS (150 g), inulin (75 g), and soy lecithin (25 g) with total 1 h of dissolution time.
4. Prehomogenize the mixture using Ultra- Turrax T25 at 1000 rpm for 5 min.

5. Add the *Chlorella* extract (150 mL, equivalent to 30 g dry extract), to achieve a core: wall ratio of 1:4 (w/w dry matter) (*see* **Note 1**).
6. High-pressure homogenize the final emulsion at 1500 bar pressure through two passes at 10 ± 2 °C inlet temperature (Table 1).
7. Measure the emulsion droplet size using dynamic light scattering (DLS, Zetasizer Nano ZS, Malvern) to achieve a mean diameter D < 2 μm and polydispersity index (PDI) < 0.3, before spray-drying [4].

3.3 Spray-Drying

1. Optimal parameters (RSM validated):
   (a) Inlet air temperature: 175 °C
   (b) Feed flow rate: 8 mL/min
   (c) Aspirator: 100%
   (d) Peristaltic pump: 20%
   (e) Outlet temperature: 85–90 °C
2. Spray the emulsion using Buchi B-290 Mini Spray dryer through 0.7 mm two- fluid nozzle (Fig. 1).

Table 1
Wall materials concentration

Material	Concentration (5 w/w dry matter)	Quantity
Maltodextrin DE 10-12	50%	250 g
Modified starch (OSA-MS)	30%	150 g
Inulin	15%	75 g
Soy lecithin	5%	25 g

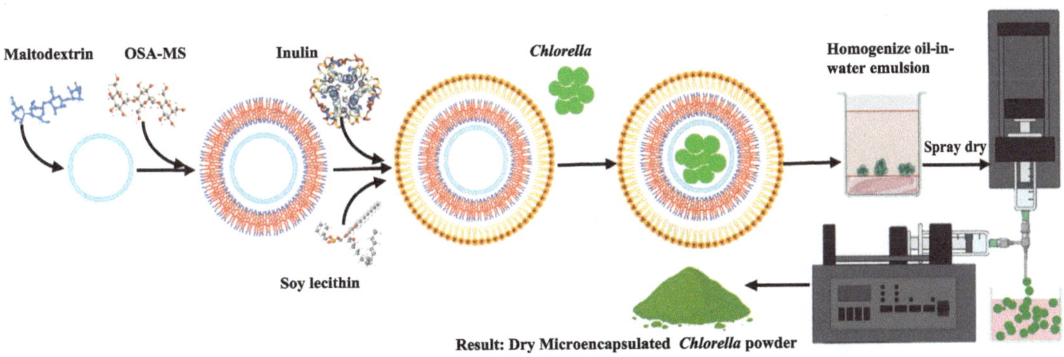

Fig. 1 Spray-dried microencapsulation of *Chlorella vulgaris* extract using multicomponent wall matrix

3. Collect the powder from the cyclone separator and immediately transfer it to a desiccator containing phosphorus pentoxide (P_2O_5) for 24 h drying at 25 °C (*see* **Note 4**).
4. Calculate the process yield as follows:

$$\text{Yield (\%)} = (\text{Mass of the collected powder}/\text{Mass of total solids in feed}) \times 100$$

5. Excepted yield range is 45–55% with powder recovery >90% from the cyclone [4].

3.4 Physiochemical Characterization of Microcapsules

3.4.1 Encapsulation Efficiency

1. Sample preparation: Accurately weigh two aliquots of powder (each ~10–50 mg)
2. Determination of surface (unencapsulated) compound:
   (a) Add 1.5 mL of cold hexane containing 0.01% Butylated Hydroxytoluene (BHT) to prevent oxidation.
   (b) Vortex the tube for 30 s, then gently invert and rotate for 2–5 min at room temperature. Do not sonicate as it may disrupt microcapsules.
   (c) Centrifuge at 6000 rpm for 5 min and collect the supernatant (surface extract) into an amber vial. If turbidity remains, wash again for 1 min. Record the volume (Vsurf).
3. Determination of total compound (complete extraction):
   (a) Add 1.5–2.0 mL of acetone:hexane mixture (1:1) and add 0.01% Butylated Hydroxytoluene (BHT).
   (b) Sonicate the mixture for 0–5 min or vortex and incubate 30 min with periodic mixing to ensure capsule disruption. Optionally, perform two sequential extractions to ensure complete disruption and combine extracts.
   (c) Centrifuge the extracts at 4000–6000 rpm for 5–10 min and collect the supernatant. Bring to known final volume (Vtot).
4. Quantification:
   (a) Analyze the extracts by HPLC (preferred) using appropriate external standards like (β-carotene, lutein, etc.) If using UV–Vis spectrophotometer measure at 450 nm and quantify with calibration curve of standard. Convert the measured concentration from mg/mL to mass per gram of powder [5].

(b) Calculation:
- Surface(mg/g) $= \frac{C_{surf} \times V_{surf}}{m}$, Total (mg/g) $= \frac{C_{tot} \times V_{tot}}{m}$
- Encapsulation Efficiency (EE %): (Total − surface/total) × 100.
- Excepted EE % = 85–92%.

3.5 Particle Morphology

1. Sample stub preparation
   (a) Attach double-sided carbon tape to a clean aluminum stub ensuring no air bubbles.
   (b) Gently sprinkle 5–10 mg of powder onto the tape using spatula and tap stub to remove loose particles.
   (c) Blow off the excess dust with compresses Nitrogen gas (15 cm distance, 5 s)
   (d) Prepare three replicate stub per sample [4].
2. Gold sputter coating
   (a) Place the mounted stubs on rotary stage of sputter chamber.
   (b) Set sputtering parameters: 20 mA sputter current, 60 s deposition time, 10-nm gold coating thickness, and initiate sputter automatic coating cycle. Verify coating uniformity by observing metallic luster under white light.
   (c) Cool stubs to room temperature for 5 min, before SEM analysis [6].
3. SEM imaging
   (a) Insert the coated stubs into 8-position sample holder, using vacuum tweezers.
   (b) Set SEM parameters (Table 2).
   (c) Capture at least 10 images per magnification per replicate.
   (d) Save as TIFF (1024 × 768, 8-bit grayscale).
   (e) Excepted morphological features: Spherical, well-formed microcapsules with smooth outer surface. Absence of collapsed or fused particle indicates adequate drying conditions. Average particle size is 10–30 μm.

Table 2
SEM parameters and purpose

Magnification	Acceleration voltage	Working distance	Spot size	Purpose
500×	5 kV	10 mm	30	Overall morphology
1000×	7 kV	9 mm	35	Surface texture
2500×	10 kV	8 mm	40	Particle fractures
5000×	10 kV	8 mm	45	Pore structure

3.6 Particle Size

1. **Instrument Setup and calibration:**
   (a) Power on Mastersizer 3000; align optics and leave15 min for warm-up.
   (b) Connect Scirocco 2000 dry feeder; set dispersive pressure 2.0 bar.
   (c) Calibrate obscuration by filling Scirocco 2000 with (5–15%) polystyrene standard (certified 20.0 μm D50, NIST traceable).
   (d) Verify D50 = 20.0 ± 0.5 μm (within ± 2.5% of certified value) and clean the funnel thoroughly using compresses air (4 bar) and vacuum suction to remove all standard residues before sample analysis.
2. **Sample Measurement**
   (a) Fill sample tray with ~1 g microencapsulated powder (avoid clumps).
   (b) Set measurement parameters:
      - Feeder speed: 50% vibration
      - Dispersive pressure: 2.0 bar
      - Measurement time: 10 s × 3 replicates
      - Obscuration target: 8–12%
      - Refractive index: 1.52 (maltodextrin matrix)
   (c) **Start measurement**: Powder auto-feeds through scirocco 2000 dry dispersion unit, a laser scans diffraction pattern and instrument software calculates the particle size distribution using diffraction pattern within 10 s per run.
3. **Data Acquisition and Validation**
   (a) Observe the obscuration trace displayed on the Mastersizer 3000 software interface during the 10-s measurement period.
   (b) Confirm obscuration stabilizes between 8–12% throughout the acquisition, indicating optimal particle concentration in the laser beam path.
   (c) Verify model fit coefficient R^2 exceeds 0.99.
   (d) Extract volume-based particle size distribution parameters from the automated software report:
      - D10: 10th percentile (10% particles smaller than this diameter)
      - D50: Median diameter (50% particles smaller, most commonly reported)
      - D90: 90th percentile (90% particles smaller than this diameter)

- Span: (D90–D10)/D50 (measure of distribution width; <1.5 = uniform)
- Sauter mean D: Surface area-weighted mean $= \Sigma(n_i \times d_i^3)/\Sigma(n_i \times d_i^2)$ (where, n_i = number of particles of diameter d_i)

(e) Repeat measurement three times using fresh powder portions from the same batch (Table 3) and calculate mean ± standard deviation [7].

(f) **Reject invalid runs exhibiting the following:**
- Obscuration <8% (insufficient sample concentration)
- Obscuration >12% (multiple scattering interference)
- $R^2 < 0.99$ (poor model fit)
- Primary beam fluctuation >±1%

(g) **Expected Valid Data**: D50 = 18.5 ± 1.2 μm, Span = 1.32 ± 0.08 (uniform, spray-dried) $R^2 = 0.995 \pm 0.003$.

3.7 Flowability Analysis

1. Weigh 10.000 ± 0.005 g microencapsulated powder using analytical balance into 50 mL graduated cylinder (Table 4).
2. Record initial volume V_0 (loose bulk density $\rho_{\text{bulk}} = \text{mass}/V_0$).

Table 3
Sample Data (3 replicate measurements of *Chlorella* microcapsules)

Replicate	D10 μm	D50 μm	D90 μm
Rep 1			
Rep 2			
Rep 3			
Mean ± SD			

Table 4
Physicochemical properties of microencapsulated *Chlorella*

Parameter	Free extract (*C. vulgaris* without any coating)	Microcapsules
Hausner ratio		
D50 (μm)		
Span		
Phenolics release 20 min (%)		

Mean ± SD ($n = 3$). $p < 0.05$

3. Place cylinder in tapped density tester and perform 1250 taps (Erweka SVM 122, 250 taps/min).
4. Record final volume V_f (tapped density $\rho_{\text{tapped}} = \text{mass}/V_f$).
5. Calculate Hausner ratio: $\text{HR} = \rho_{\text{tapped}}/\rho_{\text{bulk}}$ [3].
6. Repeat for 3 replicates. (Expected: HR 1.12–1.25 indicates excellent flowability) (*see* **Note 2**).

3.8 Dissolution Profile (Folin–Ciocalteu Method)

1. Weigh 1.000 ± 0.002 g microcapsules into 100 mL distilled water (25.0 ± 0.1 °C).
2. Place 2 cm PTFE magnetic stir bar inside the vessel and place on magnetic stirrer hotplate, with speed set at 100 ± 5 rpm.
3. At initial timing ($t = 0$) immediately withdraw 5 mL aliquot sample for total releasable phenolics content ($T = \infty$ control).
4. Withdraw 5 mL aliquots at particular time points such as 0, 5, 10, 20, 40, 60 min and replace with 5 mL fresh water.
5. Filter each aliquot through 0.45 μm PTFE syringe filter, discarding first 1 mL filtrate and dilute the filtrates (1:10) with distilled water.
6. Folin–Ciocalteu assay: pipette 0.5 mL diluted filtrate into test tubes and add 2.5 mL Folin–Ciocalteu reagent (1:10 dilution) and vortex for10 s (*see* **Note 6**).
7. Add 2 mL 7.5% Na_2CO_3 solution and vortex for 10 s
8. Incubate tubes 60 min in dark at 25 °C and measure absorbance at 765 nm against reagent blank (*see* **Note 5**).
9. Prepare gallic acid standards (0, 25, 50, 100, 250, 500 μg/mL).
10. Assay standards identically to samples (same reagents, timing).
11. Plot calibration curve: Absorbance vs. Concentration ($R^2 \geq 0.999$).
12. Interpolate concentrations (C_t, μg/mL gallic acid equivalents) from standard curve.
13. **Calculation (refer to the standard procedure)** [8]:
    (a) Absorbance vs. Concentration curve:

$$\begin{aligned}\text{Abs} &= 0.004 \times [\text{GAE}] \\ &\quad + 0.01, \text{from Absorbance calculate the concentration } C_t \\ &= (\text{Abs} - 0.01)/0.004\ [\mu\text{g/mL GAE}]\end{aligned}$$

$$\begin{aligned}\text{Example : Abs } 0.220, C_t &= (0.220 - 0.01)/0.004 \\ &= 52.5\ \mu\text{g/mL (rounded 55)}\end{aligned}$$

Table 5
Sensory scores (9-point hedonic, $n = 12$)

Attributes	Free extract	Microcapsules
Appearance		
Taste		
Odor		
Overall		
Fishy defect (1–5)		

(b) TOTAL Phenolics Released

$$\text{Total phenolics } (\mu g) = C_t \times \text{Volume correction} \times \text{Dilution factor} \times 5 \text{ mL sampled}$$

(c) $\% \text{ Release} = (\text{Total}_t/\text{Total}_{60\ \text{min}}) \times 100$

3.9 Sensory Analysis (Triangle Test + Hedonic)

1. Prepare 1% w/v dispersions (microcapsules vs. free extract) in neutral apple juice (30 mL/cup) (*see* **Note 3**).
2. Code samples with 3-digit random numbers (balanced presentation order).
3. Serve in sensory booths (22 °C, white fluorescent lighting, individual palate cleansing).
4. Triangle test ($n = 12$): Identify odd sample among 3 ($\alpha = 0.05$, $\beta = 0.10$).
5. Conduct 9-point hedonic scoring (immediately after triangle): Appearance, odor, taste, aftertaste, overall (1 = dislike extremely, 9 = like extremely). Defect intensity (1 = none, 5 = extreme): fishy odor, green color, bitterness (Table 5).
6. Analyze: χ^2 test (triangle discrimination); ANOVA + Tukey's HSD (hedonic scores) [9, 10].

4 Statistical Analysis

Perform One-way ANOVA followed by Tukey's HSD test ($p < 0.05$) using GraphPad Prism 9.0 (GraphPad Software, San Diego, CA, USA) for statistical significance.

5 Notes

1. Excessive wall material (20%+ solids) in microencapsulation increases the viscosity of the feed solution, resulting in larger

and more porous microcapsule structures, which can lead to slower and incomplete release of core bioactives during dissolution testing and reduced encapsulation efficiency. So, reduce carrier solids to 15% w/w; increase core:wall to 1:3.

2. If poor flowability is seen (Hausner ratio >1.3), sieve the microencapsulated powder through a 500 μm mesh and incorporate 2% tricalcium phosphate as an anti-caking agent to improve bulk and tapped density ratios.
3. Low Sensory Discrimination: Prepare test samples at 0.5% w/v concentration instead of 1% and screen panelists using standardized triangle tests (ISO 4121:2003) prior to hedonic evaluation.
4. Wide Particle Size Distribution (Span >2.0): Increase high-pressure homogenization passes from 2 to 3 at 1500 bar to achieve narrower droplet size distribution (D < 2 μm) before spray-drying.
5. Do not measure absorbance before 60 ± 2 min incubation in dissolution profile analysis (color unstable).
6. Do not expose Folin reaction to light (photodegradation reduces signal 15%).

References

1. Wang C-A, Onyeaka H, Miri T, Soltani F (2024) Chlorella vulgaris as a food substitute: applications and benefits in the food industry. J Food Sci. https://doi.org/10.1111/1750-3841.17529
2. Stabrauskiene J, Pudziuvelyte L, Bernatoniene J (2024) Optimizing encapsulation: comparative analysis of spray-drying and freeze-drying for sustainable recovery of bioactive compounds from Citrus × paradisi L. peels. Pharmaceuticals 17(5):596. https://doi.org/10.3390/ph17050596
3. Sansone F, Esposito T, Mencherini T, Del Prete F, Cannoniere AL, Aquino RP (2024) Exploring microencapsulation potential: multi-component spray dried delivery systems for improvement of Chlorella vulgaris extract preservation and solubility. Colloids Surf B: Biointerfaces. https://doi.org/10.1016/j.cis.2024.103297
4. Konar N, Durmaz Y, Genc Polat D, Mert B (2022) Optimization of spray drying for Chlorella vulgaris by using RSM methodology and maltodextrin. J Food Process Preserv 46(5):e16594. https://doi.org/10.1111/jfpp.16594
5. Mahfoudhi N, Hamdi S (2015) Use of almond gum and gum arabic as novel wall materials for encapsulation of flaxseed oil: impact on oxidative stability and release. Food Chem 174:227–233. https://doi.org/10.1016/j.foodchem.2014.10.142
6. Goldstein JI, Newbury DE, Michael JR, Ritchie NW, Scott JH, Joy DC (2017) Scanning electron microscopy and X-ray microanalysis, 4th edn. Springer
7. Djihad N, Fadloun Oukil N, Hamid S, Attia A, Petronilho S (2024) Microencapsulation of a green microalga (Chlorella vulgaris) by complex coacervation for its valuation as a feasible ingredient in pear snacks. Algal Res 83:103727. https://doi.org/10.1016/j.algal.2024.103727
8. Singleton VL, Orthofer R, Lamuela-Raventós RM (1999) Analysis of total phenols and other oxidation substrates and antioxidants by means of Folin-Ciocalteu reagent. Methods Enzymol 299:152–178
9. International Organization for Standardization (2003) ISO 4121:2003 sensory analysis – guidelines for the use of quantitative response scales. ISO. https://www.iso.org/standard/33817.html
10. International Organization for Standardization (2004) ISO 4120:2004. Sensory analysis – methodology – triangle test. https://www.iso.org/standard/33495.html

Chapter 36

Preparation of Bio Coating from *Chlorella* to Increase Shelf Life of Commercial Fruits

S. J. Meghana, A. Martin Paul, and D. Jayanthi

Abstract

Edible coating is a thin, invisible natural layer applied on fruits and vegetables like a protective layer that is made from natural, food-safe materials like plant extracts, algal gels, fruits pulps, seed oils, etc. The present demand from the consumers is nutritious fruits and vegetables with a longer shelf life. Bio coatings made from *Chlorella vulgaris* can be applied to various foods, also fruits and vegetables to reduce microbial growth and slow down oxidation. A number of criteria pertaining to the coating's mechanical, thermal, and physical properties can be tested. This makes them a sustainable and edible alternative to conventional packaging materials.

Key words Microalgae, *Chlorella*, Bio coating, Shelf life

1 Introduction

Edible coatings are now commonly used on many types of fruits, and they have shown very good results. When a thin, natural layer is applied on the fruit's surface, it acts like a gentle protective jacket. This layer slows down the fruit's breathing (respiration) and reduces water loss (transpiration), helping the fruit stay firm and fresh for a longer time. By controlling moisture loss and delaying softening, the coating effectively slows down the natural ageing process of the fruit, allowing it to maintain its quality during storage [1].

There has been growing interest in developing edible coatings, and researchers are actively testing different natural ingredients to improve how well they work. These coatings, made from plant-based polymers, proteins, or lipids, provide a sustainable and eco-friendly alternative to traditional packaging. Their use supports the rising demand for food preservation methods that reduce environmental impact and promote greener practices [2].

Dhanasekaran Dharumadurai (ed.), *Chlorella Genomics, Cultivation, and Bioenterprise: Green Economy through Algal Innovations*, Springer Protocols Handbooks, https://doi.org/10.1007/978-1-0716-5276-3_36,
© The Author(s), under exclusive license to Springer Science+Business Media, LLC, part of Springer Nature 2026

In the search for biobased and sustainable alternatives, researchers are increasingly investigating microalgae-derived materials for use in edible coatings. Unlike conventional coatings that rely on polysaccharides or animal-derived proteins, this study highlights *Chlorella vulgaris* as a promising microalgae-based protein source. Its sustainable origin and bioactive properties make it a valuable candidate for developing next-generation edible coating materials [3].

Blueberries, raspberries, tomatoes and other fruits and vegetables are rich in water content but are more vulnerable to external factors such as temperature irregularities and humidity accelerates deterioration, impacting both their quality and shelf life. To overcome these challenges, numerous researchers have been exploring advanced methods to enhance efficiency and effectiveness. This includes the development of edible coatings. These coatings act as a protective shield, mainly helping in freshness, increase quality, shelf life, and reducing the post-harvest losses of the fruits and vegetables [3].

2 Materials

1. *Chlorella vulgaris* pure culture.
2. Glycerol, tween 20, sodium hydroxide, ammonium sulphate.
3. Rosmarinic acid, zein, 2,2-diphenyl-1-picrylhydrazyl (DPPH).
4. Centrifuge, lyophilizer, electro spraying, spectrophotometer, penetrometer, refractometer.

3 Methods

3.1 Extraction of Proteins from Chlorella vulgaris [3]

1. Dissolve 1 g of algal culture in 18 mL of distilled water, stir and homogenous.
2. Add sodium hydroxide (1 M) dropwise with stirring till pH reaches 11.
3. Centrifuge the solution at 2550 rpm for 10 min.
4. To the supernatant, add ammonium sulphate gradually until saturation, stir for 3 h.
5. Refrigerate the mixture for at least 8 h, then centrifuge at 2550 rpm for 10 min.
6. Collect the supernatant, filter under vacuum and lyophilize the material to powder (*see* **Note 1**).

3.2 Preparation of Edible Coatings [3]

1. Prepare 6% protein solution using deionized water with continuous stirring and ensuring proteins are evenly mixed.
2. Add 50% glycerol (w/w) and tween-20, 0.4% (v/v) to enhance the flexibility, emulsify, and increase the stability of mixture.
3. Adjust the pH to 11 with 1 N NaOH.
4. Heat the mixture at 70 °C for 20 min.
5. For bioactive component encapsulation, perform electro spraying until a concentration of 5 g of rosmarinic acid per liter of coating solution is obtained.
6. Encapsulate using zein under electro spraying with a flow rate of 0.5 mL/h with voltage of 20–20.4 kV and a distance of 10 cm (*see* **Note 2**).
7. After preparation, immerse the samples in the coating solution for 2 min, air dry until the coating is completely dried on the berries.

3.3 Characterization of the Biocoating Film

Weight Loss [1–3]

1. Weigh both coated and uncoated samples every two consecutive days over a 14-day storage period using 0.1 mg weighing balance.
2. Calculate percentage of weight loss as the difference between the initial and final masses divided by initial mass.

$$\text{Weight loss\%} = (W_t - W_i / W_i)$$

3.4 Firmness [1]

1. Using a digital penetrometer with a 3 mm tip, firmness can be measured.
2. Measurements to be carried out the equatorial region, without removing the peel.
3. Maintain a penetration depth of 5 mm with a speed of the probe at 2.0 mm/s during the penetration.
4. Express the firmness in Newton (N).

Color Change [1, 3]

1. Assess the color of the samples, using a spectrophotometer with a color scale CIE L*a*b. equipped with a 4-mm measurement head.
2. Scale includes L* (lightness, from 0 = black to 100 = white), a* (red to green), and b* (yellow to blue).
3. Record each of the scales in triplicates.

Total Soluble Solids [2, 3]

1. Using a refractometer, total soluble solids are measured and expressed in °Brix.
2. Measurements is to be taken at the initial day and then twice over the entire storage period.

Total Acidity [2, 3]

1. It is measured by titration method and expressed as % citric acid.
2. Samples to be measured on initial day and then twice over the entire storage period.
3. Dilute the sample to 1:10 (v/v) with deionized water.
4. Titrate it against 0.1 M NaOH with small amount of phenolphthalein as the indicator.

$$\text{Acidity}(\%) = V_{\text{NaOH}} \cdot C_{\text{NaOH}} \cdot \text{Acidity Factor}/V_{\text{Sample}},$$

Where, V_{NaOH} = volume of NaOH consumed (mL), C_{NaOH} = concentration of NaOH (N), Acidity factor: 0.064 for citric acid and V_{Sample} = Volume of sample (mL).

Antioxidant Activity [2, 3]

1. Antioxidant activity is to be determined by the DPPH method of Brand-Williams et al. (1995).
2. Prepare DPPH solution by dissolving 2.9 mg of the active substance in 100 mL of methanol and stir at room temperature for 45 min in the absence of light.
3. Take 3.9 mL of DPPH and 0.1 mL of test sample to a cuvette and measure the absorbance at 515 nm for 20 min.
4. Observe and monitor the light absorption and decoloration of deep purple methanolic solution during the reduction reaction.
5. Calculate the free radical scavenging capacity (%RSA).

$$\%\text{RSA} = 1 - \text{AE}/\text{AD}^{*}\ 100$$

where AE is the absorption of the antioxidant solution, and AD is the absorbance of DPPH sample.

6. Calibration curve is generated by preparing various dilutions of the sample solution that relates to the concentration of the amount of DPPH remaining.
7. The remaining DPPH (DPPHrem) amount is calculated by

$$\text{DPPHrem}\ (\%) = \text{DPPH}_t/\text{DPPH}_{t=0}{}^{*}\ 100$$

The IC_{50} value (InhibitionConcentration 50%), the solution concentration at which 50% of DPPH is destroyed, is found using the calibration curve

4 Notes

1. Protein yield and effectiveness of method can be determined by Bradford method and Kjeldahl method [3, 5, 6].
2. Encapsulation process should involve an ethanol–water rosemary extract solution with a ratio of 80:20, with a zein concentration of 10% w/v. Zein is water insoluble, but dissolves in the coating solution due to higher pH of 11. While rosmarinic acid stability is known to be pH-dependent, with optimal retention observed at more neutral pH values, at pH 11, its solubility in the protein matrix was enhanced, facilitating its uniform incorporation into the coating solution. The solubility enhancement at this pH level ensured effective encapsulation and stability of the bioactive compound in the final product [3].

References

1. Medeiros Teodosio AEM, Carlos Rocha Araújo RH, Figueiredo Lima Santos BG, Linné JA, da Silva Medeiros ML, Alves Onias E, Alves de Morais F, de Melo Silva S, de Lima JF (2021) Effects of edible coatings of *Chlorella* sp. containing pomegranate seed oil on quality of *Spondias tuberosa* fruit during cold storage. Food Chem 338:127916. https://doi.org/10.1016/J.FOODCHEM.2020.127916
2. Mari A, Fafalis C, Krokida M (2024) Extension of blueberry shelf-life with edible coatings from *Chlorella vulgaris*. Chem Eng Trans 110:79–84. https://doi.org/10.3303/CET24110014
3. Mari A, Manta E, Krokida M (2025) Innovative microalgae-based edible coatings with encapsulated bioactives: enhancing fresh raspberry shelf life and quality. Processes 13:1193. https://doi.org/10.3390/pr13041193
4. Mari A, Fafalis C, Krokida M (2024) Evaluation of edible coatings from components from *Chlorella vulgaris* and comparison with conventional coatings. Coatings 14:621
5. Bradford MM (1976) A rapid and sensitive method for the quantitation of microgram quantities of protein utilizing the principle of protein-dye binding. Anal Biochem 72:248–254
6. Kjeldahl J (1883) New method for the determination of nitrogen in organic substances. SA Suppl 16:405

Chapter 37

Preparation of Bioethanol from *Chlorella* Biomass

B. R. Karthik, A. Martin Paul, D. Jayanthi, and Shreya Subramanya

Abstract

Chlorella species is one of the important feedstocks for bioethanol production due to its high carbohydrate content and growth rate. In the current protocol, *Chlorella* is cultivated under controlled laboratory conditions using nutrient media. Post harvesting, the alga is dried and subjected to chemical or enzymatic hydrolysis to convert complex polysaccharides into fermentable sugars. Fermentation of obtaining sugars by yeast leads to the production of ethanol which is further quantified using standard analytical techniques.

Key words Bioethanol, Fermentation, *Chlorella*

1 Introduction

There is a growing demand for alternative fuels due to the increasing need to reduce greenhouse gas emissions. Approximately 40% of the world's energy consumption comes from liquid fuels. Gasoline is blended with bioethanol to reduce greenhouse gas emissions. Bioethanol is biodegradable, emits fewer pollutants, and is less toxic than fossil fuels [1]. By 2050, approximately 27% of the world's transportation fuel is expected to come from bioethanol-blended biofuels [2]. Bioethanol can be produced from four types of technologies based on the feedstock used: (i) first-generation technology—uses starch and sugar-rich food crops, (ii) second-generation technology—uses lignocellulose biomass, (iii) third-generation technology—uses fast-growing algae, and (iv) fourth-generation technology—uses genetically modified microbes and algae [3, 4]. Thalloid plants such as algae, which grow in aquatic habitats, can serve as a potential source for bioethanol production. The use of algae offers several advantages, including a short life cycle, minimal or no land requirement, and no need for external feed inputs. *Chlorella* are green algae capable of photosynthesis and can accumulate large amounts of carbohydrate-rich biomass. They also have the capacity to accumulate substantial lipid

Dhanasekaran Dharumadurai (ed.), *Chlorella Genomics, Cultivation, and Bioenterprise: Green Economy through Algal Innovations*, Springer Protocols Handbooks, https://doi.org/10.1007/978-1-0716-5276-3_37,
© The Author(s), under exclusive license to Springer Science+Business Media, LLC, part of Springer Nature 2026

content, making them a promising feedstock for bioethanol production [2]. Bioethanol production from *Chlorella* involves biomass cultivation [5, 6], release of carbohydrates through chemical, enzymatic, or biological treatments, and finally the fermentation of these carbohydrates into ethanol by microbes. The present chapter describes a streamlined protocol aimed at improving the efficiency of bioethanol production.

2 Materials

1. Chu 10 medium.
2. YPD medium.
3. *Chlorella vulgaris* culture.
4. *Saccharomyces cerevisiae*.
5. Conical flask.
6. Test tubes.
7. H_2SO_4, potassium dichromate and NaOH.
8. Illumination incubator.
9. Centrifuge.
10. Sonicator.
11. Distillation system.

3 Methods

3.1 Isolation and Purification of Algae by Serial Dilution Method [7]

Composition of Chu 10 medium (mg): Ca $(NO_3)_2.4H_2O$, 2.0; $FeCl_3.6H_2O$, 0.1; H_3BO_3, 0.248; HCl (1 mol/L), 0.025 mL; KH_2PO_4, 0.62; $SO_4.7H_2O$, 2.5; $MnCl_{2.}4H_2O$, 0.139; Na_2CO_3, 2; $Na_2EDTA.2H_2O$, 0.2; $Na_2SiO_{3.}9H_2O$, 2.5; $(NH_4)_6Mo_7O_{2.}4H_2O$, 0.1; biotin, 0.1 μg; thiamine HCl, 0.1 μg; vitamin B_{12}, 1 μg; distilled water, 100 mL; pH, 6.4.

1. Ten test tubes, each containing 9 mL of Chu-10 medium is prepared.
2. One mL of the algal culture is aseptically transferred into the first tube and mixed gently to obtain a uniform suspension.
3. Subsequently, 1 mL of the suspension from the first tube is transferred into the next tube. This serial dilution is continued until the last tube.
4. All test tubes are incubated at 25 °C for two weeks (*see* **Note 1**).

3.2 Isolation and Purification of Algae by Agar Plate Method [7]

1. Chu-10 medium supplemented with agar is poured into sterile petri dishes and allowed to solidify at room temperature.
2. The algal culture is streaked onto the agar surface using the straight-line streaking method under aseptic conditions.
3. The plates are incubated in an illuminated incubator for 10–14 days under a 16:8 h light: dark cycle, while maintaining a temperature of 25 ± 2 °C.

3.3 Cultivation and Harvesting of Chlorella vulgaris [7, 8]

1. *Chlorella vulgaris* samples are obtained and inoculated into Chu-10 media at 25 ± 1 °C (*see* **Note 2**).
2. The growth flasks are maintained at room temperature (25 ± 1 °C) under a natural light: dark photoperiod of approximately 12:12 h for 25 days (*see* **Note 3**).
3. Cultures are manually shaken three times a day to ensure uniform mixing (*see* **Note 4**).
4. After 25 days of cultivation, the microalgal suspension is centrifuged, and the pellets are washed thoroughly with distilled water (*see* **Note 5**).
5. Dry the pellets at 60 °C for 48 h in a hot air over until a constant weight is achieved.
6. Mechanical grinder is used to grind dried algae into fine powder.

3.4 Carbohydrate Extraction (Pretreatment)

3.4.1 Chemical Hydrolysis (Saccharification) [4]

1. Dissolve 5 g of powdered algae in 100 mL of H_2SO_4 (1.5% or 3% (v/v)] or NaOH (*see* **Note 6**).
2. Solution is kept at ultrasonic bath for 1 h at 60 °C (Frequency of 40 kHz and Power 135 W).
3. Following hydrolysis, the samples are cooled in a water bath to room temperature, and the suspensions are centrifuged at 3000 rpm for 5 min.
4. Samples are sterilized in autoclave at 120 °C for 30 min, then filtered using cheesecloth.
5. Glucose concentrations in the supernatants are then analyzed using Colorimetric technique (Phenol–sulfuric acid method) (*see* **Note 7**).

3.4.2 Biological Hydrolysis

α-Amylase Production and Purification [8]

1. *Bacillus subtilis* and *B. cereus* are cultured and incubated at 37 °C for 48 h in a starch broth (*see* **Note 8**).
2. The culture is cold centrifuged at 10,000 rpm for 15 min at 4 °C.
3. The supernatant contains partially purified α-Amylase which is separated from the pellets which contains cell debris.
4. Starch broth is prepared by dissolving 10 g of soluble starch, 5 g of peptone, 3 g beef/yeast extract in 1 L of water (pH, 7.0).

3.5 Hydrolysis of Alga by Partially Purified α-Amylase Enzyme [8]

1. α-Amylase enzyme (200 μL) incubated with fine powdered algae for 48 h at 37 °C and the released sugar was determined using colorimetric phenol–sulfuric acid method.
2. Dried *Chlorella vulgaris* samples (60 $g{\cdot}L^{-1}$) are hydrolyzed with 0.9 mL of alpha-amylase in a 250 mL volumetric flask containing citrate buffer (pH 6.0) at the enzyme's optimal temperature of 90 °C.
3. After the addition of alpha-amylase, the mixture is cooled in a water bath until it reaches the optimal temperature for glucoamylase and cellulase (80 °C).
4. Subsequently, 2.7 mL of glucoamylase or 2.7 mL of cellulase is added, depending on the enzyme combination used (*see* **Note 9**).
5. Hydrolysis is performed at 400 rpm using a magnetic stirrer for five hours.
6. Samples are collected at 30-min intervals for glucose analysis using Colorimetric technique.

3.6 Preparation of Yeast Inoculum [1, 9]

1. 100 mL YPD medium (20 g/L, glucose; 20 g/L, peptone; 10 g/L, yeast extract) is autoclaved at 121 °C.
2. To the medium, inoculate 2 g of baker's yeast, *Saccharomyces cerevisiae,* and place in an incubation shaker at 200 rpm for 24 h at a temperature of 30 °C.

3.7 Inoculation of C. vulgaris into YPD Medium [4, 10]

1. Fermentation is carried in a 500 mL fermenter bottle or Erlenmeyer flask (close it with a cotton plug), add 200 mL of *C. vulgaris* substrate and 2.5 mL of 10% yeast solution (OD600 nm = 0.5).
2. Maintain a pH 5.5 and temperature 30 °C, also mixture is supplemented with nutrients: $(NH_4)_2SO_4$ (0.2 $g{\cdot}100\ mL^{-1}$), yeast extract (0.5 $g{\cdot}100\ mL^{-1}$), $MgSO_4$ (0.04 $g{\cdot}100\ mL^{-1}$), and KH_2PO_4 (0.1 $g{\cdot}100\ mL^{-1}$).
3. Fermenter flasks are kept in bath sonicator with 4 L of water filled in it (4.5 L total volume) and a power input of 120 W. The required temperature (30 °C) is maintained (*see* **Note 10**).
4. Incubate the mixture for 120 h (*see* **Note 11**).

3.8 Extraction of Bioethanol

3.8.1 Simple Distillation System [4]

1. Transfer the supernatant of the fermented medium to a 250 mL flask.
2. Use the simple distillation system with water as the solvent.
3. Discard the first drops of distillate and the distillation is continued for approximately 3 h.

3.9 Bioethanol Characterization

3.9.1 Analysis of Ethanol Concentration by Dichromic Acid Oxidation Method [11]

1. 4.262 g of potassium dichromate is mixed with 100 mL of distilled water using a magnetic stirrer.
2. Subsequently, add 50 mL of H_2SO_4 and make up the volume to 1000 mL by adding distilled water.
3. 10 mL of dichromic acid was gently diluted to 100 mL.
4. Pour 5 mL of dilute dichromic acid in the inner side of the Conway plate.
5. Fill 1 mL of 20% sodium carbonate and 5 mL of ethanol (sample solution) on the outer side of the Conway plate (ethanol concentration varied from 0.5 to 6 g/L).
6. Cover the Conway plate and heat in an oven at 50 °C for 2 h.
7. The absorbance values of the solution in the inner side of Conway plate was measured using a UV–Vis spectrophotometer with a maximum wavelength of 446 nm from each sample with varying ethanol concentrations to obtain a linear correlation.

4 Notes

1. Optical density of test tube sets in serial dilution is observed for progress and density of algal growth.
2. One liter of Chu-10 media that was kept in the incubator for two weeks prior to transferring the algal culture to it.
3. If the culture becomes pale, increase light or nitrate concentration. In case of contamination, repeat the procedure using subcultures.
4. Bubble aeration (using an aquarium pump) is provided for the flasks to prevent sedimentation and enhance the growth rate, if not manually shake the flasks several times in a day.
5. Wash pellets 2–3 times with Distilled water to remove salts from culture medium. Powdering algae improves surface area and better penetration of acids.
6. H_2SO_4 breaks glycosidic bonds, converts complex polysaccharides to monosaccharides, and makes sugar available for fermentation.
7. For 1 mL of sample, add 1 mL of 5% phenol solution and 5 mL conc. H_2SO_4. Keep in water bath for 25–30 °C for 20 min. Measure absorbance at 490 nm using a spectrophotometer. Carbohydrate concentration is determined by standard graph.
8. Starch slurry is prepared in cold water separately and then mixed with peptone and beef extract. Adjust pH to 7.0.
9. The variations of enzyme combinations are as follows: alpha-amylase + glucoamylase (first combination), alpha-

amylase + cellulase (second combination), and alpha-amylase + glucoamylase + cellulase (third combination).

10. Sonicator has a built-in temperate control, to verify the temperature place a thermocouple into the fermentationbroth.
11. Because of some mechanical problem, bath sonicator cannot be operated for more than 23 h. Instead, use intermitted programmed operation or built-in pulse mode.

References

1. El-Mekkawi SA, Abdo SM, Samhan FA et al (2019) Optimization of some fermentation conditions for bioethanol production from microalgae using response surface method. Bull Natl Res Cent 43:164. https://doi.org/10.1186/s42269-019-0205-8
2. Kumar VB, Pulidindi IN, Kinel-Tahan Y, Yehoshua Y, Gedanken A (2016) Evaluation of the potential of *Chlorella vulgaris* for bioethanol production. Energy Fuel 30(4): 3161–3166
3. Labbafi S, Zahedi A, Kanani B (2025) Optimizing bioethanol production from *Chlorella vulgaris* and *Dunaliella salina* using integrated renewable energy systems. Biomass Bioenergy 202:108186. https://doi.org/10.1016/J.BIOMBIOE.2025.108186
4. Silva G, Cerqueira K, Rodrigues J, Silva K, Coelho D, Souza R (2023) Cultivation of microalgae *Chlorella vulgaris* in open reactor for bioethanol production. Phycology 3(2): 325–336
5. Kim KH, Choi IS, Kim HM, Wi SG, Bae HJ (2014) Bioethanol production from the nutrient stress-induced microalga *Chlorella vulgaris* by enzymatic hydrolysis and immobilized yeast fermentation. Bioresour Technol 153:47–54. https://doi.org/10.1016/j.biortech.2013.11.059
6. Yerizam M, Jannah AM, Aprianti N (2023) Bioethanol production from Chlorella pyrenoidosa by using enzymatic hydrolysis and fermentation method. J Ecol Eng 24(1):34–40
7. Salman JM, Baiee M, Hassan F (2025) Enhancement of *Chlorella vulgaris* bioethanol production by optimizing phosphorus concentrations. J Ecol Eng 26(7):321–328
8. Hamouda RA, Sherif SA, Ghareeb MM (2018) Bioethanol production by various hydrolysis and fermentation processes with micro and macro green algae. Waste Biomass Valor 9: 1495–1501. https://doi.org/10.1007/s12649-017-9936-7
9. Megawati BZAS, Damayanti A, Putri RDA, Triwibowo B, Prasetiawan H (2022) Comparative study on the various hydrolysis and fermentation methods of *Chlorella vulgaris* biomass for the production of bioethanol. Int J Renew Energy Dev 11(2):515–522
10. Indra Neel P, Gedanken A, Schwarz R, Sendersky E (2012) Mild sonication accelerates ethanol production by yeast fermentation. Energy Fuel 26(4):2352–2356
11. Sriariyanun M, Mutrakulcharoen P, Tepaamorndech S, Cheenkachorn K, Rattanaporn K (2019) A rapid spectrophotometric method for quantitative determination of ethanol in fermentation products. Orient J Chem 35(2):744–750

Chapter 38

Preparation of Biohydrogen from *Chlorella* Biomass

Priyanka Jayam Rajendran, Muthu Roobini Rathinam, Dhanasekaran Dharumadurai, and Vishnu Raja Vijayakumar

Abstract

Biohydrogen, which is being considered as a possible replacement, is produced through the fermentation of organic substrates. This chapter focuses on the utilization of *Chlorella* sp. biomass, a high-carbohydrate microalga, as a fuel for biohydrogen produced via dark fermentation. The technique describes a methodical preprocessing strategy that effectively combines acid-thermal hydrolysis and enzymatic saccharification (using cellulase and amyloglucosidase) to convert challenging algal polysaccharides into fermentable sugars. The hydrolysate is then digested anaerobically using beneficial bacteria to optimize the amount of hydrogen produced. The possibility of a two-stage approach that integrates methanogenesis and biohydrogen synthesis is emphasized in order to significantly improve overall energy recovery. This approach provides a comprehensive framework for the conversion of microalgal biomass into clean energy, thereby promoting a carbon-neutral bioeconomy.

Key words Biohydrogen, *Chlorella*, Dark fermentation, Hydrolysate

1 Introduction

Microalgae's capacity to grow autotrophically without external organic carbon inputs, as well as their incredible photosynthetic efficiency and rapid growth rates, have made them an attractive culture for ethanol and bioenergy production. Among them, *Chlorella* species are receiving a lot of attention because of their excellent cultivation characteristics, high biomass productivity, and carbohydrate-rich cell composition, all of which make them excellent choices for fermentative biohydrogen production [1]. Their thick, multilayered cell wall, which provides hydrogen-producing microorganisms with fermentable substrates, also enables significant polysaccharide buildup.

Fossil fuels are the world's primary energy source; they are nonrenewable and have numerous negative ecological effects [2]. Consequently, various research initiatives have been established with the goal of discovering new sources of sustainable energy to

Dhanasekaran Dharumadurai (ed.), *Chlorella Genomics, Cultivation, and Bioenterprise: Green Economy through Algal Innovations*, Springer Protocols Handbooks, https://doi.org/10.1007/978-1-0716-5276-3_38,
© The Author(s), under exclusive license to Springer Science+Business Media, LLC, part of Springer Nature 2026

substitute fossil fuels. Since hydrogen only releases water when it burns, it is regarded as one of the cleanest energy sources. As a result, producing biohydrogen from microalgae has emerged as a viable technique to aid in the transition to sustainable and carbon-neutral energy.

In general, dark fermentation of microalgae for biohydrogen generation is substantially superior to photo fermentation, biophotolysis, and microbial electrolysis, because it can sequentially manufacture hydrogen and is not dependent on energy given by natural sunlight. Hydrogen synthesis by green microalgae takes place under anaerobic circumstances in the dark to trigger activation of enzymes involved in hydrogen metabolism. Since hydrogenase sensitivity to oxygen is a significant barrier to this technique, more research is needed to design a hydrogenase that is resistant to oxygen inactivation. Anaerobic conditions and a lack of sulfur encourage the growth of hydrogenases [Fe]- in algal cells, allowing for the continuous production of hydrogen [3]. Green algae produce [FeFe]-hydrogenase, an enzyme that is essential to anaerobic metabolism and becomes a more effective hydrogenase. [FeFe]-hydrogenase can catalyze the changeable oxidation of molecular hydrogen [4].

The financial sustainability of the hydrogen-manufacturing process is determined by the nature and amount of carbon; utilizing less expensive carbon sources can significantly lower hydrogen production costs. Urban wastewater can be used as a source of carbon and nutrients (N and P), and the algal cultivation system may be helpful in wastewater treatment [5]. The combination of these approaches will be the most likely commercial use in the near future and a sustainable method of creating bioenergy and bioproducts. Considering these benefits, the primary objective of this chapter is to establish whether it is possible to produce biohydrogen by using *Chlorella* biomass as a substrate for anaerobic dark fermentation. The process combines hydrogenogenic bacterial fermentation, enzymatic saccharification, and acid-thermal pretreatment to optimize hydrogen output and improve overall energy recovery.

2 Materials

1. *Chlorella* biomass.
2. Sulfuric acid (H_2SO_4) 1–2% (w/v).
3. Amyloglucosidase and cellulase enzymes.
4. Pure cultures of hydrogenogenic bacteria or heat-treated anaerobic sludge.
5. Mineral salts medium components (NaOH, HCl) for pH control.

6. Nitrogen gas (N_2) for anaerobiosis.
7. Sterile serum bottles or fermentation flasks.
8. Thermal conductivity detector (for gas analysis).

3 Method

3.1 Dark fermentation of Chlorella *Biomass*

1. **Procedure**
   - (a) Add *Chlorella* biomass at a loading rate of 20–50 g dry weight·L^{-1}.
   - (b) Mix 1% (w/v) H_2SO_4 into the biomass to reach the necessary solid–liquid ratio.
   - (c) Autoclave the mixture at 121 °C for 15–30 min to provide acid-thermal pretreatment.
   - (d) Allow the hydrolysate to cool to room temperature.
   - (e) Add amyloglucosidase and cellulase, the loading ranging from 5 to 15% (w/w of substrate) hydrolyze.
   - (f) Incubate 24–48 h at 50 °C and pH 4.8–5.5 to convert polysaccharides into fermentable sugars.
   - (g) Utilize heat-treated anaerobic sludge or active microbial cultures that produce hydrogen.
   - (h) To get free of any extra oxygen, flush the inoculum with N2 gas.
   - (i) To attain the concentration of the substrate, gradually add the inoculum.
   - (j) The hydrolysate should be transferred to sterile serum flasks or bottles with inoculum volumes of 10–30% (v/v).
   - (k) Transfer the hydrolysate to sterile serum bottles/flasks, with inoculum volumes typically 10–30% (v/v).
   - (l) Adjust pH range for hydrogenogenic pathways is 5.5–6.0.
   - (m) In order to ensure full anaerobiosis, remove the headspace with N_2 gas for five to ten minutes. Incubate at 35–37 °C (mesophilic) or 55–60 °C (thermophilic; certain investigations show thermophilic operation) (Fig. 1).
   - (n) Following treatment, utilize leftovers to produce methane in a second-stage anaerobic digester to create methane after hydrogen is produced.
   - (o) The total energy recovery and overall process efficiency are increased by this two-stage system ($H_2 \rightarrow CH_4$) (*See* **Notes 1–4**).

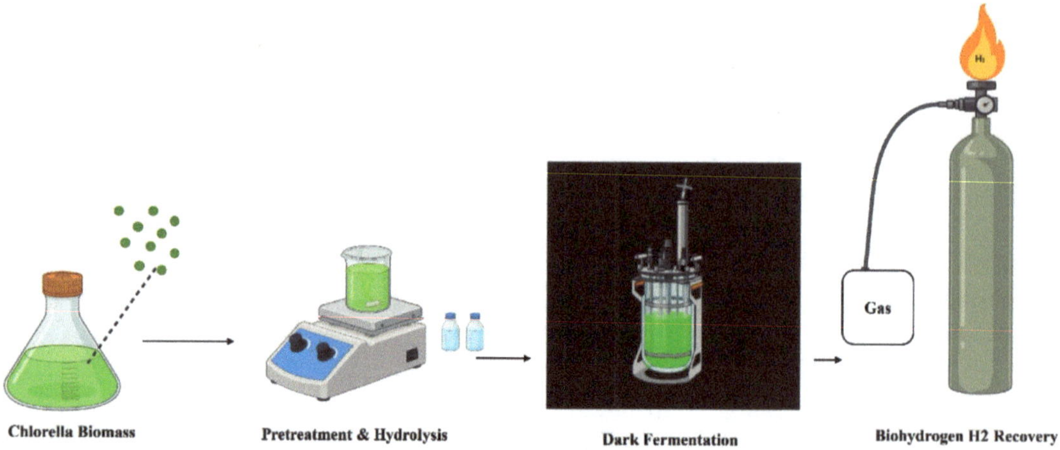

Fig. 1 Schematic representation of biohydrogen production from *Chlorella* biomass

4 Notes

1. Physical Change: After acid-thermal treatment, biomass transforms from deep green to brownish-yellow, and viscosity reduces with enzyme addition.
2. Gas Production: Visible bubbling begins within 12–24 h of incubation.
3. Maintain a pH range of 5.5–6.0 for maximum hydrogen yield.
4. To avoid splashing, add acid to water slowly during the pretreatment.

Data Availability Statement All data generated or analyzed during this study are provided in the chapter and its supplementary information files.

Ethics Approval Not applicable.

Author Contributions Conceptualization and Data collection: M.R, writing original draft, Data analyses and methodology: P.R., Supervision, Writing– review and editing: VR, DD. All contributors read the Manuscript and provided edits. All of the authors conceptualized the research. The chapter was read and approved by every author.

Competing Interests

1. *The authors declare that they have no competing financial or nonfinancial interests that could have influenced the work reported in this chapter.*

Declaration of Generative AI and AI-Assisted Technologies in the Writing Process

1. *During the preparation of this work, the author(s) used AI to improve the readability of the chapter. After using this tool, the author(s) reviewed and edited the content as needed and take (s) full responsibility for the content of the publication.*

References

1. Lay CH, Wu JH, Hsiao CL, Chang JJ, Chen CC, Lin CY (2010) Biohydrogen production from soluble condensed molasses fermentation using anaerobic fermentation. Int J Hydrog Energy 35(24):13445–13451
2. Nath K, Najafpour MM, Voloshin RA, Balaghi SE, Tyystjärvi E, Timilsina R, Eaton-Rye JJ, Tomo T, Nam HG, Nishihara H, Ramakrishna S (2015) Photobiological hydrogen production and artificial photosynthesis for clean energy: from bio to nanotechnologies. Photosynth Res 126(2):237–247
3. Stapleton JA, Swartz JR (2010) A cell-free microtiter plate screen for improved [FeFe] hydrogenases. PLoS One 5(5):e10554
4. Antal TK, Krendeleva TE, Pashchenko VZ, Rubin AB, Stensjo K, Tyystjärvi E, Ramakrishna S, Los DA, Carpentier R, Nishihara H, Allakhverdiev SI (2012) Photosynthetic hydrogen production: mechanisms and approaches. In: State of the art and progress in production of biohydrogen, pp 25–53
5. Ruiz-Marin A, Canedo-Lopez Y, Chávez-Fuentes P (2020) Biohydrogen production by *Chlorella vulgaris* and *Scenedesmus obliquus* immobilized cultivated in artificial wastewater under different light quality. AMB Express 10(1):191

Chapter 39

Fabrication of *Chlorella* Filled PVA Biocomposites via Solution Casting Method

Fayaazuddin Thajuddin, Thajuddin Nooruddin, and Dhanasekaran Dharumadurai

Abstract

Biocomposites have emerged as promising alternatives to petroleum-derived plastics due to growing environmental concerns. Among various biomass-based reinforcements, microalgae offer unique advantages, including growth, high productivity, and renewable carbon content suitable for eco-friendly composite fabrication. This chapter describes a simple solution casting protocol to fabricate eco-friendly *Chlorella*-based biocomposite films using poly(vinyl alcohol) as the main matrix. The resulting films are designed for subsequent physicochemical and mechanical characterization, providing a practical framework for developing sustainable, algae-based biocomposite materials for potential packaging and related applications.

Key words Biocomposites, Algal fillers, Poly(vinyl alcohol), *Chlorella*, Polymeric blends

1 Introduction

Significant environmental concerns related to conventional plastics drive augmenting interest in the bioplastics among the scholars and industrialists [1]. Biocomposite materials are typically formulated by two approaches where both have garnered major global interest, one employs biopolymers reinforced with organic fillers or biomass-based fillers while the other incorporates petroleum-based polymers combined with biomass-derived fillers [2]. Biocomposites generated from natural fibers, biopolymers, or bioresins represent the next generation of green composites because of their advantages such as minimal cost, weight, and their reliance on fossil resources [3]. Researchers have incorporated diverse biomass-based materials to reinforce the biocomposite encompassing micro and macroalgae, agro-industrial wastes, chitins, and celluloses [4]. The microalgae-based biomass fillers were generally made from the biomass of the strains such as *Chlorella*, *Spirulina*, and

Dhanasekaran Dharumadurai (ed.), *Chlorella Genomics, Cultivation, and Bioenterprise: Green Economy through Algal Innovations*, Springer Protocols Handbooks, https://doi.org/10.1007/978-1-0716-5276-3_39,

© The Author(s), under exclusive license to Springer Science+Business Media, LLC, part of Springer Nature 2026

Nannochloropsis to formulate the polymeric blend [5–7]. In the fabrication of composites, the biomass-based filler are combined with various polymer matrices. Among all other matrices, the poly (vinyl alcohol) (PVA) is remarkable for its favorable properties such as biodegradability, high tensile strength and water solubility, which makes them suitable for the synthesis of the composites with either organic or inorganic materials [4]. This, present chapter provides a comprehensive step-by-step protocol for the preparation of eco-friendly biocomposites using *Chlorella* sp.

2 Materials

1. Dried biomass of *Chlorella*.
2. Poly vinyl alcohol (PVA).
3. Poly diallyldimethylammonim chloride (PD).
4. Mineral oil.
5. Casting glass plate.
6. Distilled water.
7. Magnetic stirrer with heater.
8. Glass beakers.
9. Weighing balance.

3 Methodology

1. Collect the dried *Chlorella* biomass from the culture-collection center.
2. Employ solution casting technique to prepare the biocomposite material.
3. Add 2.4 g of *Chlorella* biomass in 250 mL of water in a beaker.
4. Mix vigorously by stirring for 5 min at 400 rpm.
5. Follow it by sonication using ultrasonicator.
6. Add 8 g of PVA and 4 mL of PD to the beaker containing *Chlorella* and mix vigorously for 5 min (*See* **Note 1**).
7. Now, heat the biocomposite mixture at 90 °C and keep stirring for 1 h at 350 rpm.
8. Pour the resultant heterogenous solution over a flat glass casting plate (*See* **Note 2**).
9. Allow it to dry at room temperature for 12 h [4].
10. Peel the biocomposite product from the casting plate after the incubation period (Fig. 1).
11. Store the algal biocomposite at room temperature for further characterization.

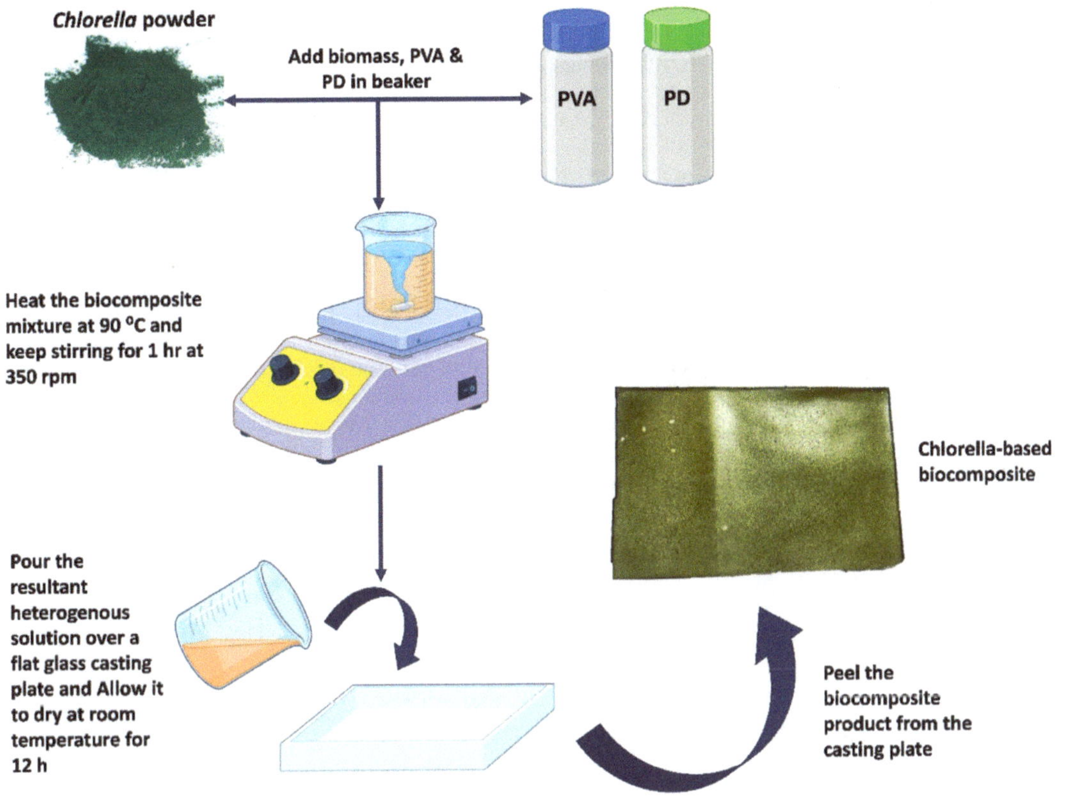

Fig. 1 Overall flow chart of *Chlorella*-based biocomposite preparation

4 Notes

1. Add PVA gradually to avoid clumping in the beaker.
2. Add mineral oil before casting the biocomposite mixture to facilitate easy removal after the incubation period.

Acknowledgments

We sincerely acknowledge the financial support from the DBT (Govt. of India), for the establishment of National Repository for Microalgae and Cyanobacteria—Freshwater & Marine (NRMC-F & M) (Phase II) (BT/PR29901/PBD26/694/2018). We gratefully acknowledge the Chief Minister Research Grant (CMRG), Govt. of Tamil Nadu for providing financial support (CMRG 2023-2024 Project No. CMRG2023BMBO7047), Rashtriya Uchchatar Shiksha Abhiyan (RUSA)–2.0 Biological Sciences (TN RUSA: 311/RUSA (2.0)/2018 Dt. December 02, 2020) and Department of Microbiology, Bharathidasan University, Tiruchirappalli, Tamil Nadu, India.

References

1. Ferreira-Filipe DA, Paço A, Duarte AC, Rocha-Santos T, Patrício Silva AL (2021) Are biobased plastics green alternatives? – a critical review. Int J Environ Res Public Health 18(15):7729
2. Faruk O, Bledzki AK, Fink HP, Sain M (2012) Biocomposites reinforced with natural fibers: 2000–2010. Prog Polym Sci 37(11): 1552–1596
3. Mukherjee T, Kao N (2011) PLA based biopolymer reinforced with natural fibre: a review. J Polym Environ 19(3):714–725
4. Tran DT, Lee HR, Jung S, Park MS, Yang JW (2018) Lipid-extracted algal biomass based biocomposites fabrication with poly (vinyl alcohol). Algal Res 31:525–533
5. Zhang F, Kabeya H, Kitagawa R, Hirotsu T, Yamashita M, Otsuki T (2000) An exploratory research of PVC-Chlorella composite material (PCCM) as effective utilization of Chlorella biologically fixing CO2. J Mater Sci 35(10): 2603–2609
6. Shi B, Wideman G, Wang JH (2012) A new approach of BioCO2 fixation by thermoplastic processing of microalgae. J Polym Environ 20(1):124–131
7. Zeller MA, Hunt R, Jones A, Sharma S (2013) Bioplastics and their thermoplastic blends from Spirulina and Chlorella microalgae. J Appl Polym Sci 130(5):3263–3275

Chapter 40

Extraction, Characterization, and Cosmetic Preparation from *Chlorella vulgaris*

G. Harita Rao, A. Martin Paul, and D. Jayanthi

Abstract

Over the years, consumers have turned to natural cosmetics as against chemical formulation because of the absence of side effects and enhanced understanding of underlying scientific principles of nature-based ingredients. *Chlorella vulgaris* is a unicellular microalga that is a rich source of chlorophyll that contains antioxidants which can act as a photo protective and anti-aging ingredient when used in cosmetic formulations. Several pigments and bioactive peptides have also been isolated from *Chlorella* spp., among other green microalga that have been observed to boost collagen production, hence paving a significant to the anti-aging class of cosmetics that have now gained popularity.

Key words *Chlorella*, Antioxidant, Chlorophyll, Cosmetics, Anti-aging

1 Introduction

Chlorella vulgaris has been a rich source of antioxidants, which have been used as functional food [1]. The global algal bioproduct market is worth more than 4 billion USD and is projected to grow to over 7 billion USD in the years to come. Microalgae are cultivated on an industrial scale for the extraction of various bioactive compounds, such as proteins, lipids, carbohydrates, and pigments that are commonly used in food additives, nutraceuticals, pharmaceuticals, therapeutics, and other high-value products [2]. Chlorophyll is a valuable bioactive compound, which is used as a natural food coloring agent and a photosensitizer for photodynamic therapy because of its antioxidant properties, anti-mutagenic ability, and near-infrared fluorescence. However, chlorophyll is highly susceptible to degradation due to changes in temperature, light, and pH [3]. Certain methods like encapsulation and spray-drying may help researchers overcome the challenges of the sensitivity of chlorophyll in related research [4–6]. An increased interest in antioxidant research has fostered interest to assess the

Dhanasekaran Dharumadurai (ed.), *Chlorella Genomics, Cultivation, and Bioenterprise: Green Economy through Algal Innovations*, Springer Protocols Handbooks, https://doi.org/10.1007/978-1-0716-5276-3_40,
© The Author(s), under exclusive license to Springer Science+Business Media, LLC, part of Springer Nature 2026

antioxidant ability of various compounds by different methods [7]. Like pigments, bioactive peptides can also be extracted from *Chlorella* spp. that can prove beneficial with its antiaging properties [8, 9]. Several phenolic compounds extracted also show photo-protective effects [10]. Exploring such properties, facemasks, sunscreens, lotions, facemasks are being made and marketed. Certain preparations are even known to promote photo-aging [11]. The chapter enumerates protocols that involve the use of bioactive compounds from *Chlorella* spp. in cosmetic formulations, their extraction, characterization, and useful effects.

2 Material

1. *Chlorella vulgaris* culture.
2. Maltodextrin, acetic acid, ethanol, potassium sulphate, copper sulphate, boric acid, sodium hydroxide, sulfuric acid, methanol, hexane, 2,2-Diphenyl-1-picrylhydrazyl reagent.
3. Ascorbic acid, methyl parabens virgin coconut oil, cetyl alcohol, propylene glycol, sodium lauryl sulphate, propylene parabenes, glycerine, triethanolamine, stearate acid, lanolin.
4. Bold's basal medium (BBM) (composition in mg/L: $NaNO_3$, 250 mg/L; K_2HPO_4, 75; $MgSO_4.7H_2O$, 75; $CaCl_2.2H_2O$, 25; KH_2PO_4, 175; NaCl, 25; alkaline EDTA solution, 1; [5 g Na_2-EDTA and 3.1 g KOH in 100 mL distilled water]; acidified iron solution, 1 mg/L [$FeSO_4.7H_2O$, 0.498 g and 0.1 mL H_2SO_4 in 100 mL distilled water]; trace metal solution, 1 mL/L [$MnCl_2.4H_2O$, 1.44 g/L; $ZnSO_4.7H_2O$, 8.82 g/L; $(NH_4)6Mo_7O_{24}.2H_2O$, 0.88 g/L; $Co(NO_3)_2.6H_2O$, 0.49 g/L; $CuSO_4.5\ H_2O$, 1.57 g/L] [12, 13].
5. BG-11 medium (composition in mg/L: $NaNO_3$, 1500; K_2HPO_4, 40; $MgSO_4.7H_2O$, 75; $CaCl_2.2H_2O$, 36; Citric acid, 6; trace metal solution, 1 mL/L [trace metal solution, $FeC_6H_5O_7.NH_4OH$, 6 g/L; Na_2-EDTA, 1 g/L; $MnCl_2.4H_2O$, 1.81 g/L; $ZnSO_4.7H_2O$, 0.222 g/L; $Na_2\ MoO_4.2H_2O$, 0.39 g/L; $CuSO_4.5H_2O$, 0.08 mg/L; H_3BO_3, 2.86 g/L] [12, 13].

3 Methods

3.1 Culturing and Extraction of C. vulgaris [1, 12–14]

1. The pure stock cultures of *Chlorella vulgaris* strains can be obtained from algal cultivation and technology institutes.
2. Media such as BBM and BG-11 are commonly used for culturing *C. vulgaris*.

3. Maintain an initial concentration of 1.0×10^6 cells/mL in medium.
4. The obtained cultures has to be maintained on BBM agar slants with a pH of 6.8 at 24 °C, under 2500 lx fluorescent light intensity and 16/8—light/dark cycle as mother culture.
5. Transfer the algal culture to 250 mL conical flask containing 125 mL of BBM liquid medium in a bioreactor.

   Or
6. The culture experiment is to be set up in 1 L photobioreactors under controlled conditions at 28 °C, with aeration of 01 vvm, photoperiod of 12:12 (hours of light:dark) using 2880 Lux fluorescent lamps with a pH of 8 on BG-11 medium.
7. Harvest cultures at regular intervals such as 7 and 10 days by natural precipitation method. Maintain bioreactor without agitation for 12 h to promote precipitation.
8. Centrifuge the *C. vulgaris* culture at 13,000–17,000 rpm to separate C. *vulgaris* from the medium.
9. The algal cells obtained is diluted with distilled water to various concentrations of 0.01; 0.05; 1 and 0.1% w/v, respectively (*see* **Note 1**).
10. Macerate the mixture with distilled water for one hour and centrifuge at 6000 rpm for 15 min to obtain filtrate and supernatant (*see* **Notes 2** and **3**).
11. The filtrate is used to obtain chlorophyll content.

3.2 Production of Chlorophyll Powder Using Maltodextrin [1]

1. Prepare the chlorophyll solution of various concentrations (0.01%, 0.05% and 0.1%) by blending with maltodextrin 100 mg/L in each solution (*see* **Note 4**).
2. Using a spray dryer, at inlet condition of 120 °C to 130 °C, outlet condition of 42 °C to 44 °C, blower 1.01 m^3/min to 1.04 m^3/min and atomizing 200 to 230 kPa to obtained chlorophyll powder (*see* **Note 5**).

3.3 Determination of Antioxidant Activity by Ascorbic Acid Equivalent Antioxidant Capacity (AEAC) [1, 15]

1. Chlorophyll powder (0.01%, 0.05% and 0.1%) is mixed with 1 mL of methanolic DPPH with vigorous stirring, homogenized, and incubated for 120 min at dark.
2. Record the absorbance at 517 nm, using distilled water as blank and ascorbic acid as standard.

3.4 Analysis of Proximate Composition

To Determine Ash Content Using Gravimetric Method (SNI 01-2891-1992) [1, 15]

1. Weigh 2 g of sample in a preweighed crucible (W_2) and char it by placing it on an electric stove.

2. Ignite the char by using a furnace set at 600 °C for 2 h until graying.
3. After graying, remove the material and cool it in a desiccator and take the final weight of crucible with material (W_1).
4. The ash content (%) is calculated based on the loss of weight before graying and after treatment (*see* **Note 6**).

$$\text{Ash content\%} = W_1/W_2{}^{*}\ 100$$

To Determine Water Content Using Gravimetric Method (SNI 01-2891-1992 Point 5.1) [1, 15]

1. Weigh 1 g of sample in a pre-weight closed bottle (W_1).
2. Drying the sample in a hot air oven at 105 °C for 3 h and cool it using desiccator for 15 min.
3. Weigh (W_2) the sample repeating until the weight becomes constant by repeating the **step 2**.

$$\text{Moisture content\%} = (W_1 - W_2)/W_1{}^{*}\ 100$$

To Determine Protein Content BY Biuret Method (Kjeltec) (SNI 01-2891-1992 Point 7.1) [16, 17]

1. Weigh 1 g of sample and transfer to a 300 mL Kjeldahl tube by adding a mixture of 2 g of selenium and 25 mL of concentrated H_2SO_4.
2. Pre heat the Kjel Digister tool and store the 300 mL Kjeldahl tube containing the sample in the tool by turning on the scrubber unit ON and destroyed.
3. Stop the KjelDigister tool after one hour and cool the tube content to reach room temperature.
4. Attach the tube to distillation unit, followed by the addition of distilled water and 40% NaOH.
5. Take 4% H_3BO_4 in a 250 mL Erlenmeyer flask as reservoir for the distillation unit.
6. Carry out distillation until the distillate volume reaches at least 3× the volume of the initial container.
7. Titrate the distillate with 0.2 N HCL solution and work on the blank was carried out every digestion cycle.

$$\text{Protein content\%} = V_1 - V_2 \times N \times 14 \times f.k./W \times 100.$$

Where:

W = weight (g) of sample; V_1 = volume of 0.001 N HCl which is used;

V_2 = volume of HCl used for titration of blank sample; N = Normality of HCl

f.k = 5.95.

To Determine Fat Content by Soxhlet Extraction Method (SNI 01-2891-1992) [16]

1. Weigh 1 g of the sample in 100 mL beaker with 25% HCl, distilled water and several boiling stones. Close the beaker and place it on hot plate.
2. Using ash paper, filter the hot residue and wash with distilled water.
3. Dry the residue in hot air oven by placing on filter paper sleeves (hulls).
4. Place the hulls in Soxhlet extractor with dry flat flask, record the weight, and add hexane.
5. Carry out the extraction using hexane.
6. Dry the fat residue in an oven at 105 °C for 30 min or until the hexane aroma/vapors are lost.
7. Cool the flask in desiccator for 15 min, weigh, and repeat drying stage until weight is constant.

$$\text{Fat content\%} = W_2 - W_1 / W_0 \times 100\%$$

Where:

W_0 = weight of sample (g)

W_1 = weight of flask before extraction with boiling stones (g)

W_2 = weight of flask with sample after extraction (g)

3.5 Calculation of Carbohydrate Percentage by Difference Method [1, 16]

Carbohydrate
= 100–(%water content + %ash content + %protein + %fat)

3.6 Production of Chlorophyll Cream [1]

1. Prepare water phase using distilled water and methyl parabens by heating in water bath at 65 °C to 70 °C.
2. Prepare oil phase with virgin coconut oil, cetyl alcohol, propylene glycol, sodium lauryl sulphate, and cetyl stearil alcohol by melting at 65 °C to 70 °C and by stirring magnetic stirrer at 200 rpm.
3. Add oil phase to water phase by stirring using a homogenizer at 200 rpm for 15 min to prevent air bubbles.

4. Obtain cream base by increasing the rpm to 800 for 10 min only when the phases are mixed well.
5. To the cream base, add chlorophyll powder, stir for another 5 min until the cream is homogeneously mixed.
6. The formulation of chlorophyll cream to be made up to following composition (%): algal extract (04%), distilled water (40%), methyl parabens (0.18%), propylene glycol (5%), cetyl stearil alcohol (4.5%), cetyl alcohol (05%), sodium lauryl sulphate (0.5%), and virgin coconut oil is added until 100%.

3.7 Production of Lotion [1]

1. Prepare water phase using distilled water, glycerine, triethanolamine, and methyl parabens by heating in water bath at 65 °C to 70 °C.
2. Prepare oil phase with stearate acid, cetyl alcohol, lanolin, and propylene parabens in water bath at 65 °C to 70 °C.
3. After the water and oil phases are homogenized, add water phase to oil phase and stir continuously to form an emulsion.
4. Add chlorophyll powder to the emulsion and homogenize completely.
5. The formulation of chlorophyll lotion to be made up to following composition (%): algal extract (03%), stearic acid (06%), methyl parabens (0.02%), lanolin (03%), triethanolamine (03%), cetyl alcohol (06%), propylene parabens (30.18%), glycerin (03%), aquadest (distilled water) is added until it reaches 150%.

4 Notes

1. Increase in concentration, yields better results. The filtrate extracted must ideally be subjected to chlorophyll extraction immediately after centrifugation as chlorophyll content in the algal extract was found to drastically reduce by any methods of storage of extract by refrigeration, freeze-drying, spray-drying, or oven-drying [2].
2. Extracting the bioactive compound by maceration prevents it from getting degraded due to heat [1]. As chlorophyll is highly sensitive to heat and light, this method of extraction is preferred [3].
3. The maceration method also commonly used because it is fast, environmentally friendly, and effectively cost, so it is feasible on an industrial scale [4].
4. Maltodextrin was used for encapsulation as increases the shelf life of the antioxidants in the chlorophyll. Maltodextrin used as an encapsulant because it has low solubility in water, high

viscosity [5]. Maltodextrin protects the antioxidant and bioactive compound release, which is induced by heat during spray-drying process [4].

5. Spray-drying method is used for chlorophyll powder production as the temperature is ideal for extracting chlorophyll without damaging the compound [6]. In spray-drying process, water evaporation occurs on a very wide surface, reducing the time for drying, increasing the thermal stability of bioactive compound because the addition of maltodextrin [4].
6. Charing of material to be done slowly. After incineration, the furnace door is to be opened slowly for the entry of oxygen. Repeatedly weigh the sample by placing in desiccator until the final weight (W_2) of the sample becomes constant [15].
7. Algal samples can be characterized for further analysis to enhance its biological significance *viz.*,
   (a) Microbial analysis–enumeration and detection for bacteria, yeast, and mold to be analyzed by total plate count method (TPC), by enumerating the microbial colonies on a nonselective agar medium under mesophilic aerobic condition [1].
   (b) Detection of *Pseudomonas aeruginosa, Candida albicans,* and *Staphylococcus aureus* are to be detected by checking the absence of microbial target growth after enrichment. Initially, the enrichment is performed using a nonselective broth medium. Later, isolation is to be performed on a selective medium followed by identification tests [1].
   (c) For morphology analysis of chlorophyll–maltodextrin rheology, Scanning Electron Microscope (SEM) can be performed on the sample by coating with gold (Au) and observing at 500× and 4000× magnifications on 20 kV voltages [1].
   (d) Amino acid analysis, vitamin determination, folates, metal ion analysis, flavor profile analysis [18].

References

1. Agustina S, Aidha N, Oktarina E (2021) The extraction of antioxidants from *Chlorella vulgaris* for cosmetics. IOP Conf Ser: Mater Sci Eng 1011
2. Silkina A, Gayo-Peláez JI, Tang KW (2025) Assessing different methods to preserve biochemical fractions in microalgal biomass for commercial applications. Algal Res 91
3. Hsiao C, Lin J, Wen H, Lin Y, Yang C (2020) Enhancement of the stability of chlorophyll using chlorophyll-encapsulated polycaprolactone microparticles based on droplet micro fluidics. Food Chem 306: 125300
4. Agustina S, Aidha NN, Oktarina E (2019) Effect of maltodextrin concentration on the characteristic of phycocyanin powder as a functional food. AIP Conf Proc 2175
5. Kurniasih RA, Purnamayati L, Amalia U, Dewi EN (2018) Formulation and characterization of phycocyanin microcapsules within maltodextrin- alginate. Agritech 38(1):23–29

6. Baldelli A, Weng Y, Prakash S, Bhandari B (2025) Optimizing microalgae bioprocessing and digestibility: spraying techniques for drying and encapsulation. Food Biosci 68
7. Philip M (2004) The use of the stable free radical diphenylpicrylhydrazyl (DPPH) for estimating antioxidant. Songklanakarin J Sci Technol 26
8. Masoumifeshani B, Abedian Kenari A, Sottorff I, Crüsemann M, Amiri Moghaddam J (2025) Identification and evaluation of antioxidant and anti-aging peptide fractions from enzymatically hydrolyzed proteins of *Spirulina platensis* and *Chlorella vulgaris*. Mar Drugs 23(4):162
9. Iriani D, Hasan B, Sari NI, Alfionita (2023) Preparation of face mask from microalga *Chlorella* sp. and its potential as antiaging. Pharmacogn J 15(1):112–118
10. Ganeson Y, Paramasivam P, Palanisamy KM, Govindan N, Maniam GP (2024) LCMS and FTIR profiling of microalga *Chlorella* sp. for cosmetics and skin care applications. Clean Water 2
11. Singh S, Purwar V (2022) Role of algae on cosmetics. Int J Creat Res Thoughts 10:73–78
12. Ilavarasi A, Mubarakali D, Praveenkumar R (2011) Optimization of various growth media to freshwater microalgae for biomass production. Biotechnology 10:540–545
13. Sharma AK, Sahoo PK, Singhal S et al (2016) Impact of various media and organic carbon sources on biofuel production potential from *Chlorella* spp. 3 Biotech 6:116. https://doi.org/10.1007/s13205-016-0434-6
14. Escobar YJB, Calambas JAM, Ramírez IDO, Muñoz-Pabon KS, Ortega JLS, Calvache JEN (2025) Production of protein biomass from Chlorella vulgaris using foliar fertilizer nutrients. Clean Waste Syst 12:100337
15. Tenda PE, Hilaria M, Wijaya H (2019) Antioxidant activities and quality test of instant faloak (Sterculia quadrifida R.Br.) from East Nusa Tenggara with added ginger (*Zingibera officinale* Roch). Int J Adv Life Sci Res 2(4):15–20. https://doi.org/10.31632/ijalsr.2019v02i04.003
16. Afifah DN, Salam DA, Nugraheni F, Resti N, Purwanti R, Anjani G, Susilo MT, Nuryanto AM, Rahmawati IS (2025) Analysis of the nutritional and fatty acid profile of sacha inchi tempe (*Plukenetia volubilis* L.) using different cooking methods. Front Nutr 12:1527865. https://doi.org/10.3389/fnut.2025.1527865
17. Junaidi SS, Rowa RH, Hamsina (2022) Nutritional content and modisco acceptance (modified dried skimmed milk and coconut oil) with additional soybeans and fruit. IOSR J Nurs Health Sci 11(1):7–12
18. Melcher F, Peest T, Garay D, Utz F, Paper M, Spaccasassi A, Obermaier L, Schneiderbanger J, Koch M, Garbe D, Nilges T, Becker T, Rychlik M, Dawid C, Brück WM, Stellner NI, Brück TB (2025) Comparative nutrient and sensory analysis of eight different commercial *Chlorella* powders. Front Food Sci Technol 5:1534438. https://doi.org/10.3389/frfst.2025.1534438

INDEX

Dhanasekaran Dharumadurai (ed.), *Chlorella Genomics, Cultivation, and Bioenterprise: Green Economy through Algal Innovations*, Springer Protocols Handbooks, https://doi.org/10.1007/978-1-0716-5276-3,

© The Editor(s) (if applicable) and The Author(s), under exclusive license to Springer Science+Business Media, LLC, part of Springer Nature 2026

C

D

G

H

I

K

L

T

U

V

W

X

Y

FSC
www.fsc.org
MIX
Papier | Fördert gute Waldnutzung
FSC® C083411

Kolibri 360 GmbH
Ferdinand-Jühlke-Straße 7
99095 Erfurt, Deutschland
produktsicherheit@kolibri360.de